NATEF Standards Job Sheets

Engine Repair (A1)

Fourth Edition

Jack Erjavec

Ken Pickerill

CENGAGE
Learning·

Australia • Brazil • Japan • Korea • Mexico • Singapore • Spain • United Kingdom • United States

CENGAGE
Learning®

NATEF Standards Job Sheets
Engine Repair (A1),
Fourth Edition
Jack Erjavec & Ken Pickerill

VP, General Manager, Skills and Planning:
Dawn Gerrain

Director, Development, Global Product
Management, Skills: Marah Bellegarde

Product Manager: Erin Brennan

Senior Product Development Manager:
Larry Main

Senior Content Developer: Meaghan Tomaso

Product Assistant: Maria Garguilo

Marketing Manager: Linda Kuper

Market Development Manager:
Jonathon Sheehan

Senior Production Director: Wendy Troeger

Production Manager: Mark Bernard

Content Project Management: S4Carlisle

Art Direction: S4Carlisle

Media Developer: Debbie Bordeaux

Cover image(s): © Snaprender/Dreamstime.com
                © Shutterstock.com/gameanna
                © i3alda/www.fotosearch.com
                © IStockPhoto.com/tarras79
                © IStockPhoto.com/tarras79
                © Laurie Entringer
                © IStockPhoto.com/Zakai
                © stoyanh/www.fotosearch.com

For product information and technology assistance, contact us at
Cengage Learning Customer & Sales Support, 1-800-354-9706

For permission to use material from this text or product,
submit all requests online at **www.cengage.com/permissions**.
Further permissions questions can be e-mailed to
permissionrequest@cengage.com

Library of Congress Control Number: 2014931169

ISBN-13: 978-1-1116-4697-4
ISBN-10: 1-111-64697-X

Cengage Learning
200 First Stamford Place, 4th Floor
Stamford, CT 06902
USA

Cengage Learning is a leading provider of customized learning solutions with
office locations around the globe, including Singapore, the United Kingdom,
Australia, Mexico, Brazil, and Japan. Locate your local office at:
www.cengage.com/global

Cengage Learning products are represented in Canada by Nelson Education, Ltd.

To learn more about Cengage Learning, visit **www.cengage.com**
Purchase any of our products at your local college store or at our preferred
online store **www.cengagebrain.com**

Notice to the Reader
Publisher does not warrant or guarantee any of the products described herein
or perform any independent analysis in connection with any of the product
information contained herein. Publisher does not assume, and expressly
disclaims, any obligation to obtain and include information other than that
provided to it by the manufacturer. The reader is expressly warned to consider
and adopt all safety precautions that might be indicated by the activities described herein and to avoid all potential hazards. By following the instructions
contained herein, the reader willingly assumes all risks in connection with such
instructions. The publisher makes no representations or warranties of any kind,
including but not limited to, the warranties of fitness for particular purpose or
merchantability, nor are any such representations implied with respect to the
material set forth herein, and the publisher takes no responsibility with respect
to such material. The publisher shall not be liable for any special, consequen-
tial, or exemplary damages resulting, in whole or part, from the readers' use of,
or reliance upon, this material.

CONTENTS

NATEF TASK LIST FOR ENGINE REPAIR

Required Supplemental Tasks (RST) List

Shop and Personal Safety

1. Identify general shop safety rules and procedures.

2. Utilize safe procedures for handling of tools and equipment.

3. Identify and use proper placement of floor jacks and jack stands.

4. Identify and use proper procedures for safe lift operation.

5. Utilize proper ventilation procedures for working within the lab/shop area.

6. Identify marked safety areas.

7. Identify the location and the types of fire extinguishers and other fire safety equipment; demonstrate knowledge of the procedures for using fire extinguishers and other fire safety equipment.

8. Identify the location and use of eyewash stations.

9. Identify the location of the posted evacuation routes.

10. Comply with the required use of safety glasses, ear protection, gloves, and shoes during lab/shop activities.

11. Identify and wear appropriate clothing for lab/shop activities.

12. Secure hair and jewelry for lab/shop activities.

13. Demonstrate awareness of the safety aspects of supplemental restraint systems (SRS), electronic brake control systems, and hybrid vehicle high-voltage circuits.

14. Demonstrate awareness of the safety aspects of high-voltage circuits (such as high intensity discharge (HID) lamps, ignition systems, injection systems, etc.).

15. Locate and demonstrate knowledge of material safety data sheets (MSDS).

Tools and Equipment

1. Identify tools and their usage in automotive applications.

2. Identify standard and metric designation.

3. Demonstrate safe handling and use of appropriate tools.

4. Demonstrate proper cleaning, storage, and maintenance of tools and equipment.

5. Demonstrate proper use of precision measuring tools (i.e., micrometer, dial-indicator, dial-caliper).

Preparing Vehicle for Service

1. Identify information needed and the service requested on a repair order.

2. Identify purpose and demonstrate proper use of fender covers, mats.

3. Demonstrate use of the three C's (concern, cause, and correction).

4. Review vehicle service history.

5. Complete work order to include customer information, vehicle identifying information, customer concern, related service history, cause, and correction.

Preparing Vehicle for Customer

1. Ensure vehicle is prepared to return to customer per school/company policy (floor mats, steering wheel cover, etc.).

Maintenance and Light Repair (MLR) Task List

A. General

A.1.	Research applicable vehicle and service information, such as internal engine operation, vehicle service history, service precautions, and technical service bulletins.	Priority Rating 1
A.2.	Verify operation of the instrument panel engine warning indicators.	Priority Rating 1
A.3.	Inspect engine assembly for fuel, oil, coolant, and other leaks; determine necessary action.	Priority Rating 1
A.4.	Install engine covers using gaskets, seals, and sealers as required.	Priority Rating 1
A.5.	Remove and replace timing belt; verify correct camshaft timing.	Priority Rating 1
A.6.	Perform common fastener and thread repair, to include: remove broken bolt, restore internal and external threads, and repair internal threads with thread insert.	Priority Rating 1
A.7.	Identify hybrid vehicle internal combustion engine service precautions.	Priority Rating 3

B. Cylinder Head and Valve Train

B.1.	Adjust valves (mechanical or hydraulic lifters).	Priority Rating 1

C. Lubrication and Cooling Systems

C.1.	Perform cooling system pressure and dye tests to identify leaks; check coolant condition and level; inspect and test radiator, pressure cap, coolant recovery tank, heater core and galley plugs; determine necessary action.	Priority Rating 1
C.2.	Inspect, replace, and adjust drive belts, tensioners, and pulleys; check pulley and belt alignment.	Priority Rating 1
C.3.	Remove, inspect, and replace thermostat and gasket/seal.	Priority Rating 1
C.4.	Inspect and test coolant; drain and recover coolant; flush and refill cooling system with recommended coolant; bleed air as required.	Priority Rating 1
C.5.	Perform engine oil and filter change.	Priority Rating 1

Automobile Service Technology (AST) Task List

A. General Engine Diagnosis; Removal and Reinstallation (R&R)

A.1.	Complete work order to include customer information, vehicle identifying information, customer concern, related service history, cause, and correction.	Priority Rating 1

A.2.   Research applicable vehicle and service information, such as internal
         engine operation, vehicle service history, service precautions,
         and technical service bulletins.   Priority Rating 1
A.3.   Verify operation of the instrument panel engine warning indicators.   Priority Rating 1
A.4.   Inspect engine assembly for fuel, oil, coolant, and other leaks;
         determine necessary action.   Priority Rating 1
A.5.   Install engine covers using gaskets, seals, and sealers as required.   Priority Rating 1
A.6.   Remove and replace timing belt; verify correct camshaft timing.   Priority Rating 1
A.7.   Perform common fastener and thread repair, to include: remove
         broken bolt, restore internal and external threads, and repair internal
         threads with thread insert.   Priority Rating 1
A.8.   Inspect, remove and replace engine mounts.   Priority Rating 2
A.9.   Identify hybrid vehicle internal combustion engine service precautions.   Priority Rating 3
A.10. Remove and reinstall engine in an OBDII or newer vehicle; reconnect
         all attaching components and restore the vehicle to running condition.   Priority Rating 3

B. Cylinder Head and Valve Train Diagnosis and Repair
B.1.   Remove cylinder head; inspect gasket condition; install cylinder head
         and gasket; tighten according to manufacturer's specifications
         and procedures.   Priority Rating 1
B.2.   Clean and visually inspect a cylinder head for cracks; check gasket
         surface areas for warpage and surface finish; check passage condition.   Priority Rating 1
B.3.   Inspect pushrods, rocker arms, rocker arm pivots and shafts for wear,
         bending, cracks, looseness, and blocked oil passages (orifices);
         determine necessary action.   Priority Rating 2
B.4.   Adjust valves (mechanical or hydraulic lifters).   Priority Rating 1
B.5.   Inspect and replace camshaft and drive belt/chain; includes checking
         drive gear wear and backlash, end play, sprocket and chain wear,
         overhead cam drive sprocket(s), drive belt(s), belt tension, tensioners,
         camshaft reluctor ring/tone-wheel, and valve timing components;
         verify correct camshaft timing.   Priority Rating 1
B.6.   Establish camshaft position sensor indexing.   Priority Rating 1

C. Engine Block Assembly Diagnosis and Repair
C.1.   Remove, inspect, or replace crankshaft vibration damper
         (harmonic balancer).   Priority Rating 2

D. Lubrication and Cooling Systems Diagnosis and Repair
D.1.   Perform cooling system pressure and dye tests to identify leaks; check
         coolant condition and level; inspect and test radiator, pressure cap,
         coolant recovery tank, heater core and galley plugs; determine
         necessary action.   Priority Rating 1
D.2.   Identify causes of engine overheating.   Priority Rating 1
D.3.   Inspect, replace, and adjust drive belts, tensioners, and pulleys;
         check pulley and belt alignment.   Priority Rating 1
D.4.   Inspect and test coolant; drain and recover coolant; flush and refill
         cooling system with recommended coolant; bleed air as required.   Priority Rating 1
D.5.   Inspect, remove, and replace water pump.   Priority Rating 2
D.6.   Remove and replace radiator.   Priority Rating 2
D.7.   Remove, inspect, and replace thermostat and gasket/seal.   Priority Rating 1
D.8.   Inspect and test fan(s) (electrical or mechanical), fan clutch,
         fan shroud, and air dams.   Priority Rating 1
D.9.   Perform oil pressure tests; determine necessary action.   Priority Rating 1
D.10. Perform engine oil and filter change.   Priority Rating 1

D.11.	Inspect auxiliary coolers; determine necessary action.	Priority Rating 3
D.12.	Inspect, test, and replace oil temperature and pressure switches and sensors.	Priority Rating 2

Master Automobile Service Technology (MAST) Task List

A. General Engine Diagnosis; Removal and Reinstallation (R&R)

A.1.	Complete work order to include customer information, vehicle identifying information, customer concern, related service history, cause, and correction.	Priority Rating 1
A.2.	Research applicable vehicle and service information, such as internal engine operation, vehicle service history, service precautions, and technical service bulletins.	Priority Rating 1
A.3.	Verify operation of the instrument panel engine warning indicators.	Priority Rating 1
A.4.	Inspect engine assembly for fuel, oil, coolant, and other leaks; determine necessary action.	Priority Rating 1
A.5.	Install engine covers using gaskets, seals, and sealers as required.	Priority Rating 1
A.6.	Remove and replace timing belt; verify correct camshaft timing.	Priority Rating 1
A.7.	Perform common fastener and thread repair, to include: remove broken bolt, restore internal and external threads, and repair internal threads with thread insert.	Priority Rating 1
A.8.	Inspect, remove, and replace engine mounts.	Priority Rating 2
A.9.	Identify hybrid vehicle internal combustion engine service precautions.	Priority Rating 3
A.10.	Remove and reinstall engine in an OBDII or newer vehicle; reconnect all attaching components and restore the vehicle to running condition.	Priority Rating 3

B. Cylinder Head and Valve Train Diagnosis and Repair

B.1.	Remove cylinder head; inspect gasket condition; install cylinder head and gasket; tighten according to manufacturer's specifications and procedures.	Priority Rating 1
B.2.	Clean and visually inspect a cylinder head for cracks; check gasket surface areas for warpage and surface finish; check passage condition.	Priority Rating 1
B.3.	Inspect pushrods, rocker arms, rocker arm pivots and shafts for wear, bending, cracks, looseness, and blocked oil passages (orifices); determine necessary action.	Priority Rating 2
B.4.	Adjust valves (mechanical or hydraulic lifters).	Priority Rating 1
B.5.	Inspect and replace camshaft and drive belt/chain; includes checking drive gear wear and backlash, end play, sprocket and chain wear, overhead cam drive sprocket(s), drive belt(s), belt tension, tensioners, camshaft reluctor ring/tone-wheel, and valve timing components; verify correct camshaft timing.	Priority Rating 1
B.6.	Establish camshaft position sensor indexing.	Priority Rating 1
B.7.	Inspect valve springs for squareness and free height comparison; determine necessary action.	Priority Rating 3
B.8.	Replace valve stem seals on an assembled engine; inspect valve spring retainers, locks/keepers, and valve lock/keeper grooves; determine necessary action.	Priority Rating 3
B.9.	Inspect valve guides for wear; check valve stem-to-guide clearance; determine necessary action.	Priority Rating 3
B.10.	Inspect valves and valve seats; determine necessary action.	Priority Rating 3
B.11.	Check valve spring assembled height and valve stem height; determine necessary action.	Priority Rating 3
B.12.	Inspect valve lifters; determine necessary action.	Priority Rating 2

B.13.   Inspect and/or measure camshaft for runout, journal wear
and lobe wear.   Priority Rating 2

B.14.   Inspect camshaft bearing surface for wear, damage, out-of-round,
and alignment; determine necessary action.   Priority Rating 3

C. Engine Block Assembly Diagnosis and Repair

C.1.   Remove, inspect, or replace crankshaft vibration damper
(harmonic balancer).   Priority Rating 2

C.2.   Disassemble engine block; clean and prepare components for
inspection and reassembly.   Priority Rating 1

C.3.   Inspect engine block for visible cracks, passage condition, core
and gallery plug condition, and surface warpage; determine
necessary action.   Priority Rating 2

C.4.   Inspect and measure cylinder walls/sleeves for damage, wear,
and ridges; determine necessary action.   Priority Rating 2

C.5.   Deglaze and clean cylinder walls.   Priority Rating 2

C.6.   Inspect and measure camshaft bearings for wear, damage,
out-of-round, and alignment; determine necessary action.   Priority Rating 3

C.7.   Inspect crankshaft for straightness, journal damage,
keyway damage, thrust flange and sealing surface condition,
and visual surface cracks; check oil passage condition; measure end
play and journal wear; check crankshaft position sensor reluctor ring
(where applicable); determine necessary action.   Priority Rating 1

C.8.   Inspect main and connecting rod bearings for damage and wear;
determine necessary action.   Priority Rating 2

C.9.   Identify piston and bearing wear patterns that indicate connecting
rod alignment and main bearing bore problems; determine
necessary action.   Priority Rating 3

C.10. Inspect and measure piston skirts and ring lands; determine
necessary action.   Priority Rating 2

C.11. Determine piston-to-bore clearance.   Priority Rating 2

C.12. Inspect, measure, and install piston rings.   Priority Rating 2

C.13. Inspect auxiliary shaft(s) (balance, intermediate, idler,
counterbalance or silencer); inspect shaft(s) and support bearings
for damage and wear; determine necessary action; reinstall and time.   Priority Rating 2

C.14. Assemble engine block.   Priority Rating 1

D. Lubrication and Cooling Systems Diagnosis and Repair

D.1.   Perform cooling system pressure and dye tests to identify leaks;
check coolant condition and level; inspect and test radiator,
pressure cap, coolant recovery tank, heater core and galley plugs;
determine necessary action.   Priority Rating 1

D.2.   Identify causes of engine overheating.   Priority Rating 1

D.3.   Inspect, replace, and adjust drive belts, tensioners, and pulleys;
check pulley and belt alignment.   Priority Rating 1

D.4.   Inspect and test coolant; drain and recover coolant; flush and refill
cooling system with recommended coolant; bleed air as required.   Priority Rating 1

D.5.   Inspect, remove, and replace water pump.   Priority Rating 2

D.6.   Remove and replace radiator.   Priority Rating 2

D.7.   Remove, inspect, and replace thermostat and gasket/seal.   Priority Rating 1

D.8.   Inspect and test fan(s) (electrical or mechanical), fan clutch,
fan shroud, and air dams.   Priority Rating 1

D.9.   Perform oil pressure tests; determine necessary action.   Priority Rating 1

D.10. Perform engine oil and filter change.   Priority Rating 1
D.11. Inspect auxiliary coolers; determine necessary action.   Priority Rating 3
D.12. Inspect, test, and replace oil temperature and pressure
         switches and sensors.   Priority Rating 2
D.13. Inspect oil pump gears or rotors, housing, pressure relief devices,
         and pump drive; perform necessary action.   Priority Rating 2

DEFINITION OF TERMS USED IN THE TASK LIST

To clarify the intent of these tasks, NATEF has defined some of the terms used in the task listings. To get a good understanding of what the task includes, refer to this glossary while reading the task list.

add	To increase fluid or pressure to the correct level or amount.
adjust	To bring components to specified operational settings.
assemble (reassemble)	To fit together the components of a device or system.
bleed	To remove air from a closed system.
check	To verify condition by performing an operational or comparative examination.
clean	To rid components of foreign matter for the purpose of reconditioning, repairing, measuring, and reassembling.
concentricity	A comparison of the center point between circular measurements in relation to each other.
confirm	To acknowledge something has happened with firm assurance.
deglaze	To remove a smooth glossy surface.
demonstrate	To show or exhibit the knowledge of a theory or procedure.
determine	To establish the procedure to be used to perform the necessary repair.
determine necessary action	Indicates that the diagnostic routine(s) is the primary emphasis of a task. The student is required to perform the diagnostic steps and communicate the diagnostic outcomes and corrective actions required, addressing the concern or problem. The training program determines the communication method (worksheet, test, verbal communication, or other means deemed appropriate) and whether the corrective procedures for these tasks are actually performed.
diagnose	To identify the cause of a problem.
differentiate	To perceive the difference in or between one thing to other things.
disassemble	To separate a component's parts as a preparation for cleaning, inspection, or service.
drain	To use gravity to empty a container.
fill (refill)	To bring the fluid level to a specified point or volume.
flush	To internally clean a component or system.
hone	To restore or resize a bore by using rotating cutting stones.
identify	To establish the identity of a vehicle or component prior to service; to determine the nature or degree of a problem.
inspect	(see *check*)
install (reinstall)	To place a component in its proper position in a system.
maintain	To keep something at a specified level, position, rate, etc.
measure	To determine existing dimensions/values for comparison to specifications.
on-board diagnostics (OBD)	Diagnostic protocol that monitors computer inputs and outputs for failures.

perform	To accomplish a procedure in accordance with established methods and standards.
perform necessary action	Indicates that the student is to perform the diagnostic routine(s) and perform the corrective action item. Where various scenarios (conditions or situations) are presented in a single task, at least one of the scenarios must be accomplished.
pressure test	To use air or fluid pressure to determine the integrity, condition, or operation of a component or system.
priority ratings	Indicates the minimum percentage of tasks, by area, that a program must include in its curriculum in order to be certified in that area.
reassemble	(see *assemble*)
refill	(see *fill*)
remove	To disconnect and separate a component from a system.
repair	To restore a malfunctioning component or system to operating condition.
replace	To exchange a component; to reinstall a component.
research	To make a thorough investigation into a situation or matter.
select	To choose the correct part or setting during assembly or adjustment.
service	To perform a procedure as specified in the owner's or service manual.
test	To verify a condition through the use of meters, gauges, or instruments.
torque	To tighten a fastener to a specified degree of tightness (in a given order or pattern if multiple fasteners are involved on a single component).
vacuum test	To determine the integrity and operation of a vacuum- (negative pressure) operated component and/or system.
verify	To confirm that a problem exists after hearing the customer's complaint or concern, or to confirm the effectiveness of a repair.

CROSS-REFERENCE GUIDES

MLR Task	Job Sheet
A.6	15
A.7	17
B.1	22
C.1	42
C.2	44
C.3	48
C.4	42
C.5	50

AST Task	Job Sheet
A.1	10
A.2	10
A.3	11
A.4	12
A.5	13
A.6	14
A.7	15
A.8	16
A.9	17
A.10	18, 19
B.1	20
B.2	20
B.3	21
B.4	22
B.5	23
B.6	23
C.1	31, 41
D.1	42
D.2	43
D.3	44
D.4	42
D.5	45
D.6	46
D.7	47
D.8	48
D.9	49
D.10	50
D.11	51
D.12	52

MAST Task	Job Sheet
A.1	10
A.2	10
A.3	11
A.4	12
A.5	13
A.6	14
A.7	15
A.8	16
A.9	17
A.10	18, 19

MAST Task	**Job Sheet**
B.1	20
B.2	20
B.3	21
B.4	22
B.5	23
B.6	23
B.7	24
B.8	25
B.9	26
B.10	27
B.11	28
B.12	29
B.13	30
B.14	30
C.1	31, 41
C.2	32
C.3	33
C.4	34
C.5	35
C.6	36
C.7	37
C.8	38
C.9	38
C.10	39
C.11	39
C.12	39
C.13	40
C.14	41
D.1	42
D.2	43
D.3	44
D.4	42
D.5	45
D.6	46
D.7	47
D.8	48
D.9	49
D.10	50
D.11	51
D.12	52
D.13	53

PREFACE

The automotive service industry continues to change with the technological changes made by automobile, tool, and equipment manufacturers. Today's automotive technician must have a thorough knowledge of automotive systems and components, good computer skills, exceptional communication skills, good reasoning, the ability to read and follow instructions, and above average mechanical aptitude and manual dexterity.

This new edition, like the last, was designed to give students a chance to develop the same skills and gain the same knowledge that today's successful technician has. This edition also reflects the changes in the guidelines established by the National Automotive Technicians Education Foundation (NATEF), in 2013.

The purpose of NATEF is to evaluate technician training programs against standards developed by the automotive industry and recommend qualifying programs for certification (accreditation) by ASE (National Institute for Automotive Service Excellence). Programs can earn ASE certification upon the recommendation of NATEF. NATEF's national standards reflect the skills that students must master. ASE certification through NATEF evaluation ensures that certified training programs meet or exceed industry-recognized, uniform standards of excellence.

At the expense of much time and thought, NATEF has assembled a list of basic tasks for each of their certification areas. These tasks identify the basic skills and knowledge levels that competent technicians have. The tasks also identify what is required for a student to start a successful career as a technician. In June 2013, after many discussions with the industry, NATEF established a new model for automobile program standards. This new model is reflected in this edition and covers the new standards that are based on three (3) levels: Maintenance & Light Repair (MLR), Automobile Service Technician (AST), and Master Automobile Service Technician (MAST). Each successive level includes all the tasks of the previous level in addition to new tasks. In other words, the AST task list includes all of the MLR tasks plus additional tasks. The MAST task list includes all of AST tasks plus additional tasks specifically for MAST.

Most of the content in this book are job sheets. These job sheets relate to the tasks specified by NATEF, according to the appropriate certification level. The main considerations during the creation of these job sheets were student learning and program certification by NATEF. Students are guided through standard industry-accepted procedures. While they are progressing, they are asked to report their findings as well as offer their thoughts on the steps they have just completed. The questions asked of the students are thought provoking and require students to apply what they know to what they observe.

The job sheets were also designed to be generic. That is, whenever possible, the tasks can be performed on any vehicle from any manufacturer. Also, completion of the sheets does not require the use of specific brands of tools and equipment. Rather, students use what is available. In addition, the job sheets can be used as a supplement to any good textbook.

Also included are description and basic use of the tools and equipment listed in NATEF's standards. The standards recognize that not all programs have the same needs, nor do all programs teach all of the NATEF tasks. Therefore, the basic philosophy for the tools and equipment requirement is that the training should be as thorough as possible with the tools and equipment necessary for those tasks.

Theory instruction and hands-on experience of the basic tasks provide initial training for employment in automotive service or further training in any or all of the specialty areas. Competency in the tasks indicates to employers that you are skilled in that area. You need to know the appropriate theory, safety, and support information for each required task. This should include identification and use of the required tools and testing and measurement equipment required for the tasks, the use of current reference and training materials, the proper way to write work orders and warranty reports, and the storage, handling, and use of hazardous materials as required by the "Right to Know" law, and federal, state, and local governments.

Words to the Instructor: We suggest you grade these job sheets based on completion and reasoning. Make sure the students answer all the questions. Then, look at their reasoning to see if the task was actually completed and to get a feel for their understanding of the topic. It will be easy for students to copy others' measurements and findings, but each student should have their own base of understanding and that will be reflected in their explanations.

Words to the Student: While completing the job sheets, you have a chance to develop the skills you need to be successful. When asked for your thoughts or opinions, think about what you observed. Think about what could have caused those results or conditions. You are not being asked to give accurate explanations for everything you do or observe. You are only asked to think. Thinking leads to understanding. Good technicians are good because they have a basic understanding of what they are doing and why they are doing it.

Jack Erjavec and Ken Pickerill

ENGINES

Before you complete the job sheets, some basics must be covered. This discussion begins with an overview of engines. Emphasis is placed on what they do and how they work. This includes the major components and designs of engines and their role in the efficient operation of engines of all designs.

Preparation for working on an automobile is not complete if certain safety issues are not first addressed. A discussion of safety covers those things you should and should not do when working on engines. Included are proper ways to deal with hazardous and toxic materials.

NATEF's task list for Engine Repair certification is also given with definitions of some of the terms used to describe the tasks. This list gives you a good look at what the experts say you need to know before you can be considered competent to work on engines.

Following the task list are descriptions of the various tools and types of equipment you need to be familiar with. These are the tools you will use to complete the job sheets. They are also the tools NATEF has identified as being necessary for servicing engines.

Following the tool discussion is a cross-reference guide that shows what NATEF tasks are related to specific job sheets. In most cases there are single job sheets for each task. Some tasks are part of a procedure and when this occurs one job sheet may cover two or more tasks. The remainder of the book contains the job sheets.

BASIC ENGINE THEORY

The engine provides the power to drive the wheels of the vehicle. All gasoline and diesel automobile engines are classified as internal combustion engines because the combustion or burning that creates energy takes place inside the engine. Combustion is the burning of an air and fuel mixture. As a result of combustion, large amounts of pressure are generated in the engine. This pressure or energy is used to power the vehicle. The engine must be built strong enough to hold the pressure and temperatures formed by combustion.

Four-Stroke Cycle

Nearly all automotive engines run through the four-stroke cycle to produce power (Figure 1). These strokes or events repeat themselves, in each cylinder, several times per minute. The name of the stroke defines what event takes place at each stroke. There is the intake stroke during which air and fuel are pushed into the cylinder. The compression stroke compresses the air and fuel mixture to prepare it for ignition. The power stroke is the one that supplies the power. It occurs after ignition and results from the quick expansion of the air/fuel mixture during combustion.

Engine Identification

Before you begin any work on an engine you need to make a positive identification of the engine. It is best to do this by locating the Vehicle Identification Number (VIN). The VIN is visible through the windshield on the driver's side of the vehicle. The coding used in the VIN can tell you much about the vehicle, including what engine the vehicle was originally manufactured with.

Cylinder Block

The biggest part of the engine is the cylinder block. The cylinder block is a large casting of metal that is bored with holes to allow for the

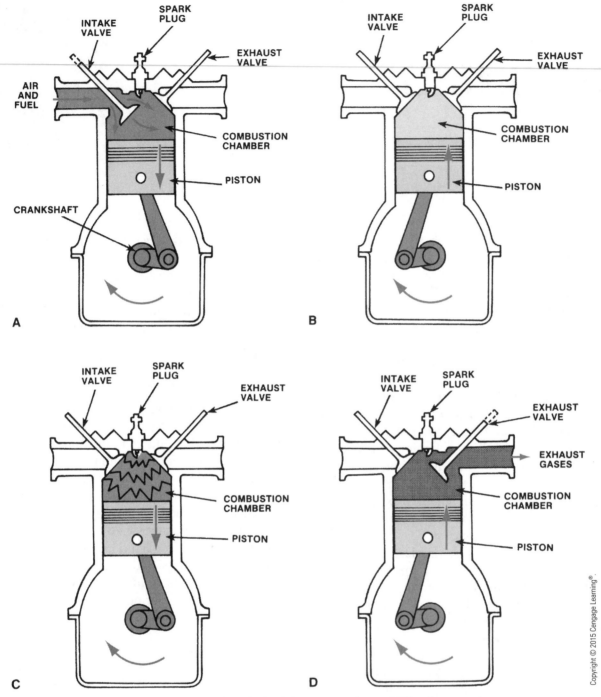

Figure 1 (A) Intake stroke, (B) compression stroke, (C) power stroke, and (D) exhaust stroke.

passage of lubricants and coolant through the block and provide spaces for movement of mechanical parts. The block contains the cylinders, which are round passageways fitted with pistons. Combustion takes place in the cylinders as the pistons move through their four-stroke cycle. A cylinder block is normally one piece, cast, and machined so that all of the parts contained in it fit properly. Some late-model engines are two-piece castings bolted together.

Blocks are typically made of iron or aluminum. Aluminum blocks normally have steel sleeves inserted into the cylinder bores. These provide a stronger, more durable surface for the pistons to move up and down in.

The cylinder bores are critical areas of service for a technician. The walls of the bore must be the correct size and be without flaws in order to allow the piston to move smoothly up and down the bore while it maintains a seal against the walls.

Pistons

The burning of air and fuel takes place between the cylinder head and the top of the piston. The piston is a can-shaped part closely fitted inside the cylinder. In a four-stroke cycle engine, the piston moves through four different movements or strokes to complete one cycle. On the intake stroke, the piston moves downward, and a charge of air/fuel mixture is introduced into the cylinder. As the piston travels upward, the air/fuel mixture is compressed in preparation for burning. Just before the piston reaches the top of the cylinder, ignition occurs and combustion starts. The pressure of expanding gases forces the piston downward on its power stroke. When it reciprocates, or moves upward again, the piston is on the exhaust stroke. During the exhaust stroke, the piston pushes the burned gases out of the cylinder.

The gap between the outside of the piston and the cylinder walls is sealed by piston rings. These rings of steel are formed under tension so that they expand outward from the piston to the cylinder walls to form a good seal.

If the piston doesn't seal well, the efficiency of its four strokes decreases. Consider the intake stroke. The piston moves down to increase the volume of the cylinder and to lower the pressure in the cylinder. If there is leakage around the piston rings, air from the crankcase, which is below the piston, will enter the cylinder. This in turn will lower the amount of vacuum formed and reduce the amount of air/fuel mixture that will be drawn into the cylinder. The other strokes are affected in the same way.

Making sure the piston is able to move freely up and down the cylinder while maintaining a seal is a critical task during an engine overhaul.

Connecting Rods and Crankshaft

The up and down motion of the pistons must be converted to rotary motion before it can drive the wheels of a vehicle. This conversion is achieved by linking the piston to a crankshaft with a connecting rod. The upper end of the connecting rod moves with the piston. The lower end of the connecting rod is attached to the crankshaft and moves in a circle. The end of the crankshaft is connected to the transmission to continue the power flow through the drive train and to the wheels.

The crankshaft is housed by the engine block and is held in place by main bearing caps. Between the main bearing caps and the crankshaft and the cylinder block and the crankshaft are flat insert bearings.

The crankshaft does not rotate directly on the bearings; instead, it rotates on a film of oil trapped between the bearing surface and the crankshaft journals. The area that holds the oil film is critical to the life of an engine. If the crankshaft journals become out-of-round, tapered, or scored, or if the bearings are worn, the proper oil film will not form. Without the correct oil film, premature wear will occur on the bearings and crankshaft journals. This can also result in crankshaft breakage. Therefore, another critical area for technicians is the one between the crankshaft journals and the bearing surfaces.

Cylinder Head

The cylinder head mounts on top of the engine block and serves as the uppermost seal for the combustion chamber. To aid in that sealing, a head gasket is sandwiched between the cylinder head and engine block. The combustion chamber is an area into which the air/fuel mixture is compressed and burned. The cylinder head contains all or most of the combustion chamber. The cylinder head also contains ports, which are passageways through which the air/fuel mixture enters and burned gases exit the cylinder. The head also contains the valves, which must open to allow air/fuel in and allow exhaust gases out. When the valves are closed, they too need to provide a positive seal. The cylinder head may also be the mounting spot or serve as the housing for a camshaft.

Much of the work done by a technician on a cylinder head is done to ensure a good seal. A good seal is formed between the deck of the block and the cylinder head when both surfaces are flat and parallel to each other. This not only provides a seal for the combustion chamber but also allows the oil and coolant passages in the head to positively connect with passages in the engine block.

Valves

All engines have at least two valves: an intake and an exhaust. A valve is a movable part that opens and closes a passageway. When the valves open, intake air flows into the combustion chamber, or exhaust gases leave it. When closed, they seal the chamber. The heads of the intake and exhaust valves have different diameters. The intake valve is the larger of the two. An exhaust valve can be smaller because exhaust gases move more easily than intake air.

Most newer engines use multivalve arrangements. Multivalve engines can have three, four, or five valves per cylinder. The most common arrangement is four valves per cylinder, with two intake and two exhaust valves. Using additional intake and exhaust valves results in a more complete combustion, which reduces the chances of misfire and detonation. It also results in better fuel efficiency, a cleaner exhaust, and increased power output.

An engine's valves close into a valve seat. The interference fit between the face of the valve and the valve seat provide for the seal. Since this seal is made by two metal surfaces, it is critical that the surfaces be smooth and perfectly round.

The valves move in a valve guide that is typically pressed into the cylinder head. The clearance between the stem of the valve and the inside diameter of the guide provides some lubrication for the valve's movement. If the clearance is excessive or irregular, the valve may become canted while it moves and may not provide for a good seal when it is closed. Excessive valve to guide clearance will also allow oil to enter into the combustion chamber from the valve train. To prevent excessive oil from passing through the guides, oil seals are installed on the valve guides or valve stems.

A valve has two primary purposes: it must open and it must close. The opening of the valve is controlled directly or indirectly by a camshaft. A camshaft controls the movement of the valves, causing them to open and close at the proper time. The closing of the valve is the primary job of a valve spring. The valve spring forces the valve closed when the camshaft no longer is putting pressure on the valve.

The valve spring is held in place by a spring retainer and valve locks or keepers. This assembly also works to hold the valve in place in the cylinder head.

Camshafts

All current engines have at least one camshaft; some have more. Vee-type engines with dual overhead camshafts have four. Regardless of the number of camshafts an engine has, there is one camshaft lobe for each valve. That lobe is responsible for the opening and closing of that valve. The rotation of a camshaft is timed with the crankshaft, at half the speed of the crank. The camshaft and crankshaft must be perfectly in time with each other in order for the engine to efficiently pass through the four strokes of the engine's pistons.

The sides of a camshaft lobe have ramps that control how quickly the valve opens and closes. The highest point of the lobe gives the maximum valve opening. When the valve is fully open, more air/fuel can enter the cylinder or more exhaust gas can exit the cylinder. The base circle of the lobe provides no valve lift and the valve is closed.

Camshafts rotate on insert bearings, similar to those used with the crankshaft. Camshaft bearings for OHV engines and some OHC engines are full-round insert bearings. The bearings are pressed into bores in the engine block or cylinder head. Many OHC engines use split bearings. These engines have bearing caps that secure the camshaft in the cylinder head or in a camshaft housing mounted to the top of the cylinder head.

Meshed gears or a timing chain and gears can drive the camshaft in an OHV engine. The camshafts of OHC engines are either driven by timing belts and sprockets, or timing chains and gears. Most belts and chain drives have a tensioner to maintain the proper tightness. OHC engines with a belt drive system have sprockets on the crankshaft and the camshaft, which are linked by a continuous neoprene belt. The belt has square-shaped internal teeth that mesh with teeth on the sprockets.

The sprockets on the camshaft and the crankshaft can be linked by a continuous chain. Nearly all OHV engines use a chain drive system. Late-model OHC engines may have many drive chains. Those with one chain use it to drive the camshaft(s) by the crankshaft. Some have a long chain to drive the camshafts and a short chain to drive the balance shaft. Other engines have an additional chain that aligns the intake camshaft with the exhaust camshaft. Vee-type engines may have a separate chain from the crankshaft to the camshafts of each cylinder bank, and then additional chains to connect the intake and exhaust camshafts on each bank.

Valve Train

The valve train consists of a combination of these parts: lifter, pushrod, and rocker arm. OHV engines use all three to transfer the movement of the camshaft lobes to the valves. On OHC engines, the camshaft may directly or indirectly move the valves. When the camshaft directly controls the valves, the camshaft is placed above the valves and the lobes ride on the top of the valves. Often, the camshaft indirectly drives the valves. These setups utilize a compact setup with a rocker arm and lifter.

It is important that the correct clearance between the top of the valve and the mechanism operating the valve be maintained. The gap allows for the expansion of parts caused by heat. If the gap is too small, the expanded size of the parts may cause the valves to remain slightly open when they should be closed. The advantage of hydraulic lifters is that they compensate for the expansion and do not require periodic adjustment. All mechanical lifters require valve lash adjustments. The method used to adjust valve lash varies from valve train to valve train. The procedure may be simply adjusting a screw or it may involve replacing a shim.

Variable Valve Timing

Normally, the timing and opening (lift) of the valves are controlled by fixed lobes on a camshaft. Engineers design these lobes to meet the engine's needs at the anticipated normal operating speeds and loads, and to achieve maximum fuel economy and minimum emission levels during those operating conditions.

Changing valve timing in response to driving conditions improves driveability and lowers fuel consumption and emission levels. Many different variable valve timing (VVT) systems are used in today's engines. Many systems only vary the timing of the intake or exhaust valves, some vary the timing of both valves, others vary the lift and timing of the intake or exhaust valves, and a few vary the timing and lift of both valves. An ECM advances or retards valve timing based on operating conditions, such as engine speed and load.

VVT systems are either staged or continuously variable designs. Most staged valve timing systems switch between two or more different camshaft profiles based on operating conditions. This can be done by using a camshaft that has different lobes for each of the intake valves, with a rocker arm over each of the lobes. Each lobe is designed for a different operating condition and engine speed. When the appropriate condition and speed is met, the ECM activates the correct rocker arm, through oil pressure, so it opens and closes the valves according to the profile of the cam lobe it rides on.

Continuously variable systems change the phasing or timing of a valve's duration (open and close times). These systems provide a wider torque curve, reduction in fuel consumption, improved power at high speeds, and a reduction in hydrocarbon and NO_x emissions. To provide continuously variable value timing, camshafts are fitted with a phaser. The phaser is mounted where a timing pulley, sprocket, or gear would be. Phasers are either based on a helical gear set or vanes enclosed in a housing. They can be electronically or hydraulically controlled.

In a hydraulically controlled system, oil flow is controlled by the ECM. Electronic systems rely on stepper motors. In some engines, the phaser shifts the timing of the intake cam lobes independently of the exhaust cam lobes and changes only the valve timing of the intakes.

A unique setup for variably controlling intake and exhaust valve timing and lift relies on a conventional camshaft, ground for high performance, with ultra-high-speed valves to bleed off the fluid in the hydraulic lifters. Based on inputs to the PCM, the fluid pressure in the lifter is changed to delay valve openings, change the amount of time the valve is opened, or prevent valves from opening at all, which makes it possible to shut down cylinders until more power is needed. A solenoid is used to control the flow of engine oil into a piston in each lifter that effectively determines the tappet's height.

Intake and Exhaust Systems

A manifold is metal ductwork assembly used to direct the flow of gases to or from the combustion chambers. Two separate manifolds are attached to the cylinder head. The intake manifold delivers a mixture of air and fuel to the intake ports. The exhaust manifold mounts over the exhaust ports and carries exhaust gases away from the cylinders.

The intake and exhaust systems are critical to the efficient operation of an engine. An engine needs air to operate. This air is drawn into the cylinders through the intake or intake induction system during the intake stroke of the pistons. This air is then mixed with fuel and delivered to the combustion chambers.

After combustion, burnt gases are present in the combustion chamber. In order to make room for a fresh supply of intake air and fuel, those exhaust gases must leave the combustion chamber. The exhaust system has that responsibility.

Although there seems to be an endless supply of air outside the engine, often not enough air is available for the engine. Because the actual time of the intake stroke is so short, there is little time to gather, direct, and force air into the cylinders. A vehicle's air induction system is designed to move as much air as possible and as quietly as possible. This statement is more true for high performance vehicles than normal vehicles, simply because engine performance is directly proportional to intake airflow.

A typical intake system consists of an air cleaner assembly, air filter, air ductwork, and intake manifold. These components are designed to quickly deliver a fresh and clean supply of air to the engine's cylinders. The design of the components determines the maximum amount of air that can be delivered and the actual amount is determined by the position of the engine's throttles plates and the vacuum formed on the pistons' intake stroke.

Many engines have been equipped with turbochargers and superchargers to increase the amount and speed of air delivery to the engine. A supercharger is driven by the crankshaft with a drive belt. A turbocharger uses the flow of exhaust gases and is not mechanically connected to the engine. Both of these devices are designed to force pressurized air into the combustion chambers during their intake strokes. Some late-model engines have two turbochargers, one for low speed and another for high engine speeds.

A highly efficient intake system is worthless if the exhaust gases cannot leave the combustion chamber. In order for the cylinders to form a strong vacuum or low pressure, the cylinders need to be empty. The exhaust system has this responsibility. Once the exhaust has left the cylinder, a strong vacuum can be formed and a fresh supply of intake air and fuel can enter the cylinder. This allows for the beginning of another four-stroke cycle.

As the piston moves up on its exhaust stroke, the exhaust gases are pushed out into the exhaust system. From there, the gases must travel quite a distance to leave the exhaust system and be released into the atmosphere. To be able to exit the exhaust system, the exhaust gases must remain under pressure and be hotter than the outside air. As the exhaust moves through the exhaust system, it cools some and the pressure drops. The flow of exhaust will always move from a point of higher pressure to a point of lower pressure. Therefore, as long as the pressure right in front of the moving exhaust gases is lower than the pressure of the exhaust, the exhaust will continue to flow out. An efficient exhaust system ensures that this happens.

Exhaust systems normally face one of two problems: leaks or restrictions. Leaks not only cause excessive noise, but can also decrease the temperature and pressure of the exhaust. Restrictions cause exhaust pressure to build up before the restriction. The difference in pressure determines the rate of flow. When the pressure difference is slight, less exhaust will be able to leave the cylinders and there will be less room for the fresh intake charge.

Lubrication System

The moving parts of an engine need constant lubrication. Lubrication limits the amount of wear and reduces the amount of friction in the engine. Friction occurs when two objects rub against each other, generating heat. The engine relies on the circulation of engine oil to keep things healthy and to minimize wear.

Oil is delivered under pressure to all moving parts of the engine. Engine damage will occur if dirt gets trapped in small clearances designed for an oil film or if dirt blocks an oil passage. The oil circulating through the engine is filtered by an oil filter that should be replaced whenever the oil is changed. Periodic oil and filter changes ensure that clean oil circulates throughout the engine.

A screen on the oil pickup tube initially filters the engine's oil. The oil pump draws oil out of the oil pan through the pickup tube. The oil pan serves as a reservoir for the oil. Oil drawn from the pan is circulated through the engine. After circulating, the oil drips back into the oil pan.

The oil pump is driven by the crankshaft or camshaft. The pump is responsible for the circulation of pressurized oil throughout the engine. Since an oil pump is a positive displacement pump, an oil pressure relief valve is needed to prevent excessively high oil pressures. Pressure is controlled to make sure the oil will flow over the parts, not spray over the parts.

The pressurized oil travels through oil passages or galleries in the engine, crankshaft, connecting rods, and other parts.

Problems with the engine's lubrication system will cause engine problems. Sometimes lubrication problems are not obvious, and at other times (such as when there are oil leaks) they are very obvious. Likewise, sometimes the causes of these problems can be quite obvious, and at other times not so obvious. Diagnosis of the system can be challenging.

Cooling System

The burning of the air/fuel mixture in the combustion chambers of the engine produces large amounts of heat. This heat must not be allowed to build up and must be reduced, because it can easily damage and warp the metal parts of an engine. To remove this heat, a heat-absorbing liquid, called engine coolant, circulates inside the engine. The system responsible for the circulation of coolant and the dissipation of heat is called the engine's cooling system.

A typical cooling system relies on a water pump that circulates the coolant through the system. The engine typically drives the pump. The coolant is a mixture of water and antifreeze. The coolant is pushed through passages, called water jackets, in the cylinder block and head to remove heat from the area around the cylinders' combustion chambers. The heat picked up by the coolant is sent to the radiator. The radiator transfers the coolant's heat to the outside air as the coolant flows through its tubes. To help remove the heat from the coolant, a cooling fan is used to pull cool outside air through the fins of the radiator. The cooled liquid is then returned to the engine to repeat the cycle.

Since parts of the cooling system are located in various spots under the vehicle's hood, hoses are used to connect these parts and keep the system sealed.

A thermostat is used to control the temperature by controlling the amount of coolant that moves into the radiator. When the engine is cold, the thermostat is closed and no coolant flows to the radiator. This allows the engine to warm up to the correct operating temperature. As the coolant warms, the thermostat opens and allows coolant to flow through the radiator. The hotter the coolant gets, the wider the thermostat is open.

Engine coolant is actually a mixture of antifreeze/coolant and water. This mixture lowers the freezing point of the coolant and raises its boiling point. The pressure of the system also increases the boiling point of the coolant. The cooling system is a closed system. The pressure in the system is the result of the operation of the water pump and the increase in temperature. As temperature increases, so does pressure.

To raise the boiling point of the coolant even further, the cooling system is pressurized. The radiator pressure cap maintains this pressure. The cap is designed to keep the cooling system sealed until a particular pressure is reached. At that time, the cap allows some of the pressure to be vented from the system. This action prevents excessive cooling system pressures.

Not only can cooling system problems affect the durability of an engine, they can also cause many driveability problems. Today's computer controls are set to keep the engine operating at the best temperature for efficiency. This efficiency results in lower emissions and improved driveability. Problems in the cooling and lubrication systems can cause the engine to run at higher than normal temperatures; likewise some problems in the cooling system can cause the engine to run cooler than normal.

Atkinson Cycle Engines

An Atkinson cycle engine is a four-stroke cycle engine in which the intake valve is held open longer than normal during the compression stroke. This is done in most engines, and is used in hybrid vehicles with a VVT system.

As the piston is moving up, the mixture is being compressed and some of it pushed back into the intake manifold. As a result, the amount of mixture in the cylinder and the engine's effective displacement and compression ratio are reduced. Often the Atkinson cycle is referred to as a five-stroke cycle because there are two distinct cycles during the compression stroke. The first is while the intake valve is open and the second is when the intake valve is closed. This two-stage compression stroke creates the "fifth" cycle.

In a conventional engine, much engine power is lost due the energy required to compress the mixture during the compression stroke. The Atkinson cycle reduces this power loss and this leads to greater engine efficiency. The Atkinson cycle also effectively changes the length of the time the mixture is being compressed. Most Atkinson cycle engines have a long piston stroke. Keeping the intake valve open during compression effectively shortens the stroke. However since the valves are closed during the power stroke, that stroke is long. The longer power stroke allows the combustion gases to expand more and reduces the amount of heat that is lost during the exhaust stroke. As a result, the engine runs more efficiently than a conventional engine.

Although these engines provide improved fuel economy and lower emissions, they also produce less power. The lower power results from the lower operating displacement and compression ratio. Power is lower also because these engines take in less air than a conventional engine.

Hybrid Vehicles

A hybrid vehicle has at least two different types of power or propulsion systems. Today's hybrid vehicles have an internal combustion engine and an electric motor (Some vehicles have more than one electric motor.). A hybrid's electric motor is powered by batteries and/or ultra-capacitors, which are recharged by a generator that is driven by the engine. They are also recharged through regenerative braking. The engine may use gasoline, diesel, or an alternative fuel. Complex electronic controls monitor the operation of the vehicle. Based on the current operating conditions, electronics control the engine, electric motor, and generator.

Depending on the design of the hybrid vehicle, the engine may power the vehicle, assist the electric motor while it is propelling the vehicle, or drive a generator to charge the vehicle's batteries. The electric motor may propel the vehicle by itself, assist the engine while it is propelling the vehicle, or act as a generator to charge the batteries. Many hybrids rely exclusively on the electric motor(s) during slow speed operation, the engine at higher speeds, and both during some certain driving conditions.

Many hybrid vehicles have Atkinson cycle engines. The low power output from the engine is supplemented with the power from the electric motors. This combination offers good fuel economy, low emissions, and normal acceleration.

Diesel Engines

Diesel engines are the dominant power plant in heavy-duty trucks, construction equipment, farm equipment, buses, and marine applications. Diesel engines in cars and light trucks will become more common soon.

The operation of a diesel engine is comparable to a gasoline engine. They also have a number of components in common, such as the crankshaft, pistons, valves, camshaft, and water and oil pumps. They both are available as four-stroke combustion cycle engines. However, diesel engines have compression ignition systems. Rather than relying on a spark for ignition, a diesel engine uses the heat produced by compressing air in the combustion chamber to ignite the fuel. The compression ratio of diesel engines is typically three times (as high as 25:1) that of a gasoline engine. As intake air is compressed, its temperature rises. Just before the air is fully compressed, a fuel injector sprays a small amount of diesel fuel into the cylinder. The high temperature of the compressed air instantly ignites the fuel. The combustion causes increased heat in the cylinder and the resulting high pressure moves the piston down on its power stroke.

Diesel engines are heavier than gasoline engines of the same power. A diesel engine must be made stronger to contain the extremely high compression and combustion pressures. A diesel engine, also, produces less horsepower than a same sized gasoline engine. Therefore to provide the required power, the displacement of the engine is increased. This results in a physically larger engine. Diesels have high torque outputs at very low engine speeds but do not run well at high engine speeds.

Today's automotive diesel engines are considered "clean diesels." This is due to the cleaner fuel used by them and the various systems (injection and emissions) incorporated into the engines' systems.

SAFETY

In an automotive repair shop, there is great potential for serious accidents, simply because of the nature of the business and the equipment used. When people are careless, the automotive repair industry can be one of the most dangerous occupations. But, the chances of your being injured when working on a car are close to nil if you learn to work safely and use common sense. Shop safety is the responsibility of everyone in the shop.

Personal Protection

Some procedures, such as grinding, result in tiny particles of metal and dust that are thrown off at very high speeds. These metal and dirt particles can easily get into your eyes, causing scratches or cuts on your eyeball. Pressurized gases and liquids escaping a ruptured hose or hose-fitting can spray a great distance. If these chemicals get into your eyes, they can cause blindness. Dirt and sharp bits of corroded metal can easily fall down into your eyes while you are working under a vehicle.

Eye protection should be worn whenever you are exposed to these risks. To be safe, you should wear safety glasses whenever you are working in the shop. Some procedures may require that you wear other eye protection in addition to safety glasses. For example, when cleaning parts with a pressurized spray, you should wear a face shield. The face shield not only gives added protection to your eyes but also protects the rest of your face.

If chemicals such as battery acid, fuel, or solvents get into your eyes, flush them continuously with clean water. Have someone call a doctor and get medical help immediately.

Your clothing should be well fitted and comfortable but made of strong material. Loose, baggy clothing can easily get caught in moving parts and machinery. Some technicians prefer to wear coveralls or shop coats to protect their personal clothing. Your work clothing should offer you some protection but should not restrict your movement.

Long hair and loose, hanging jewelry can create the same type of hazard as loose-fitting clothing. They can get caught in moving engine parts and machinery. If you have long hair, tie it back or tuck it under a cap.

Never wear rings, watches, bracelets, and neck chains. These can easily get caught in moving parts and cause serious injury.

Always wear shoes or boots of leather or similar material with non-slip soles. Steel-tipped safety shoes can give added protection to your feet. Jogging or basketball shoes, street shoes, and sandals are inappropriate in the shop.

Good hand protection is often overlooked. A scrape, cut, or burn can limit your effectiveness at work for many days. A well-fitted pair of heavy work gloves should be worn during operations such as grinding and welding or when handling hot components. Always wear approved rubber gloves when handling strong and dangerous caustic chemicals.

Many technicians wear thin, surgical-type latex gloves whenever they are working on vehicles. These offer little protection against cuts but do offer protection against disease and grease buildup under and around your fingernails. These gloves are comfortable and are quite inexpensive.

Accidents can be prevented simply by the way you act. The following are some guidelines to follow while working in a shop. This list does not include everything you should or shouldn't do; it merely presents some things to think about.

- Never smoke while working on a vehicle or while working with any machine in the shop.

- Playing around is not fun when it sends someone to the hospital. Such things as air nozzle fights, creeper races, and practical jokes have no place in the shop.

- To prevent serious burns, keep your skin away from hot metal parts such as the radiator, exhaust manifold, tailpipe, catalytic converter, and muffler.

- Always disconnect electric engine cooling fans when working around the radiator. Many of these will turn on without warning and can easily chop off a finger or hand. Make sure you reconnect the fan after you have completed your repairs.

- When working with a hydraulic press, make sure the pressure is applied in a safe manner. It is generally wise to stand to the side when operating the press.

- Properly store all parts and tools by putting them away in a place where people will not trip over them. This practice not only cuts down on injuries, but also reduces time wasted looking for a misplaced part or tool.

Work Area Safety

Your entire work area should be kept clean and safe. Any oil, coolant, or grease on the floor can make it slippery. To clean up oil, use a commercial oil absorbent. Keep all water off the floor. Water not only makes smooth floors slippery, but it also is dangerous as a conductor of electricity.

Aisles and walkways should be kept clean and wide enough to allow easy movement. Make sure the work areas around machines are large enough so that the machinery can be safely operated.

Gasoline is a highly flammable volatile liquid. Something that is flammable catches fire and burns easily. A volatile liquid is one that vaporizes very quickly. Flammable volatile liquids are potential firebombs. Always keep gasoline or diesel fuel in an approved safety can and never use gasoline to clean your hands or tools.

Handle all solvents (or any liquids) with care to avoid spillage. Keep all solvent containers closed, except when pouring. Proper ventilation is very important in areas where volatile solvents and chemicals are used. Solvents and other combustible materials must be stored in approved and designated storage cabinets or rooms with adequate ventilation. Never light matches or smoke near flammable solvents and chemicals, including battery acids.

Oily rags should also be stored in an approved metal container. When these oily, greasy, or paint-soaked rags are left lying about or are not stored properly, they can cause spontaneous combustion. Spontaneous combustion results in a fire that starts by itself, without a match.

Disconnecting the vehicle's battery before working on the electrical system, or before welding, can prevent fires caused by a vehicle's electrical system. To disconnect the battery, remove the negative or ground cable from the battery and position it away from the battery.

Hybrid vehicles have special disconnect points for the high-voltage battery. BEFORE starting work, be sure to put on any special protective gear that needs to be worn, such as high-voltage gloves and boots. Technicians working on hybrids are required to attend special training prior to working on a live vehicle.

Know where all of the shop's fire extinguishers are located. Fire extinguishers are clearly labeled as to what type they are and what types of fire they should be used on. Make sure you use the correct type of extinguisher for the type of fire you are dealing with. A multipurpose dry chemical fire extinguisher will put out ordinary combustibles, flammable liquids, and electrical fires. Never put water on a gasoline fire because that will just cause the fire to spread. The proper fire extinguisher will smother the flames.

During a fire, never open doors or windows unless it is absolutely necessary; the extra draft will only make the fire worse. Make sure the fire department is contacted before or during your attempt to extinguish a fire.

Tool and Equipment Safety

Careless use of simple hand tools such as wrenches, screwdrivers, and hammers causes many shop accidents that could be prevented. Keep all hand tools grease-free and in good condition. Tools that slip can cause cuts and bruises. If a tool slips and falls into a moving part, it can fly out and cause serious injury.

Use the proper tool for the job. Make sure the tool is of professional quality. Using poorly made tools or the wrong tools can damage parts or the tool itself, or could cause injury. Never use broken or damaged tools.

Safety around power tools is very important. Serious injury can result from carelessness. Always wear safety glasses when using power tools. If the tool is electrically powered, make sure it is properly grounded. Before using it, check the wiring for cracks in the insulation, as well as for bare wires. Also, when using electrical power tools, never stand on a wet or damp floor. Never leave a running power tool unattended.

Tools that use compressed air are called pneumatic tools. Compressed air is used to inflate tires, apply paint, and drive tools. Compressed air can be dangerous when it is not used properly.

When using compressed air, safety glasses and/or a face shield should be worn. Particles of dirt and pieces of metal, blown by the high-pressure air, can penetrate your skin or get into your eyes.

Before using a compressed air tool, check all hose connections. Always hold an air nozzle or air control device securely when starting or shutting off the compressed air. A loose nozzle can whip suddenly and cause serious injury. Never point an air nozzle at anyone. Never use compressed air to blow dirt from your clothes or hair. Never use compressed air to clean the floor or workbench.

Always be careful when raising a vehicle on a lift or a hoist. Adapters and hoist plates must be positioned correctly to prevent damage to the underbody of the vehicle. There are specific lift points that allow the weight of the vehicle to be evenly supported by the adapters or hoist plates. The correct lift points can be found in the vehicle's service information. Before operating any lift or hoist, carefully read the operating manual and follow the operating instructions.

Once you feel the lift supports are properly positioned under the vehicle, raise the lift until the supports contact the vehicle. Then, check the supports to make sure they are in full contact with the vehicle. Shake the vehicle to make sure it is securely balanced on the lift, and then raise the lift to the desired working height. Before working under a car, make sure the lift's locking devices are engaged.

A vehicle can be raised off the ground by a hydraulic jack. The jack's lifting pad must be positioned under an area of the vehicle's frame or at one of the manufacturer's recommended lift points. Never place the pad under the floor pan or under steering and suspension components, because these are easily damaged by the weight of the vehicle. Always position the jack so the wheels of the vehicle can roll as the vehicle is being raised.

Safety stands, also called jack stands, should be placed under a sturdy chassis member, such as the frame or axle housing, to support the vehicle after it has been raised by a jack. Once the safety stands are in position, the hydraulic pressure in the jack should be slowly released until the weight of the vehicle is on the stands. Never move under a vehicle when it is supported only by a hydraulic jack. Rest the vehicle on the safety stands before moving under the vehicle.

Heavy parts of the automobile, such as engines, are removed with chain hoists or cranes. Cranes often are called cherry pickers. To prevent serious injury, chain hoists and cranes must be properly attached to the parts being lifted. Always use bolts with enough strength to support the object being lifted. After you have attached the lifting chain or cable to the part that is being removed, have your instructor check it. Place the chain hoist or crane directly over the assembly, and then attach the chain or cable to the hoist.

Parts cleaning is a necessary step in most repair procedures. Always wear the appropriate protection when using chemical, abrasive, and thermal cleaners.

Vehicle Operation

When the customer brings a vehicle in for service, certain driving rules should be followed to ensure your safety and the safety of those working around you. For example, before moving a car into the shop, buckle your safety belt. Make sure no one is near, the way is clear, and there are no tools or parts under the car before you start the engine. Check the brakes before putting the vehicle in gear. Then, drive slowly and carefully in and around the shop.

If the engine must be running while you are working on the car, block the wheels to prevent the car from moving. Place the transmission into park for automatic transmissions or into neutral for manual transmissions. Set the parking (emergency) brake. Never stand directly in front of or behind a running vehicle.

Run the engine only in a well-ventilated area to avoid the danger of poisonous carbon monoxide (CO) in the engine exhaust. CO is an odorless but deadly gas. Most shops have an exhaust ventilation system, and you should always use it. Connect the hose from the vehicle's tailpipe to the intake for the vent system. Make sure the vent system is turned on before running the engine. If the work area does not have an exhaust venting system, use a hose to direct the exhaust out of the building.

Working Safely on High-Voltage Systems

Electric drive vehicles (battery-operated, hybrid, and fuel cell electric vehicles) have high-voltage electrical systems (from 42 volts to 650 volts). These high voltages and amperages can kill you! Fortunately, most high-voltage circuits are identifiable by size and color. The cables have thicker insulation and are typically colored orange. The connectors are also colored orange. On some vehicles, the high-voltage cables are enclosed in an orange shielding or casing, again the orange indicates high voltage. In addition, the high-voltage battery pack and most high-voltage components have "High Voltage" caution labels. Be careful not to touch these wires and parts.

Wear insulating gloves, commonly called "lineman's gloves," when working on or around the high-voltage system. These gloves must be class "0" rubber insulating gloves, rated at 1000-volts. Also, to protect the integrity of the insulating gloves, as well as you, wear leather gloves over the insulating gloves while doing a service.

Make sure they have no tears, holes or cracks and that they are dry. Electrons are very small and can enter through the smallest of holes in your gloves. The integrity of the gloves should be checked before using them. To check the condition of the gloves, blow enough air into each one so they balloon out. Then fold the open end over to seal the air in. Continue to slowly fold that end of the glove toward the fingers. This will compress the air. If the glove continues to balloon as the air is compressed, it has no leaks. If any air leaks out, the glove should be discarded. All gloves, new and old, should be checked before they are used.

There are other safety precautions that should always be adhered to when working on an electric drive vehicle:

- Always adhere to the safety guidelines given by the vehicle's manufacture.
- Obtain the necessary training before working on these vehicles.
- Be sure to perform each repair operation following the test procedures defined by the manufacturer.
- Disable or disconnect the high-voltage system before performing services to those systems. Do this according to the procedures given by the manufacturer.
- Anytime the engine is running in a hybrid vehicle, the generator is producing high voltage and care must be taken to prevent being shocked.
- Before doing any service to an electric drive vehicle, make sure the power to the electric motor is disconnected or disabled.
- Systems may have a high-voltage capacitor that must be discharged after the high-voltage system has been isolated. Make sure to wait the prescribed amount of time (normally about 10 minutes) before working on or around the high-voltage system.
- After removing a high-voltage cable, cover the terminal with vinyl electrical tape.
- Always use insulated tools.
- Alert other technicians that you are working on the high-voltage systems with a warning sign such as "High-Voltage Work: Do Not Touch!"
- Always install the correct type of circuit protection device into a high-voltage circuit.

- Many electric motors have a strong permanent magnet in them; individuals with a pacemaker should not handle these parts.
- When an electric drive vehicle needs to be towed into the shop for repairs, make sure it is not towed on its drive wheels. Doing this will drive the generator(s), which can overcharge the batteries and cause them to explode.
- Always tow these vehicles with the drive wheels off the ground or move them on a flat bed.

HAZARDOUS MATERIALS AND WASTES

A typical shop contains many potential health hazards for those working in it. These hazards can cause injury, sickness, health impairments, discomfort, and even death. Here is a short list of the different classes of hazards:

- Chemical hazards are caused by high concentrations of vapors, gases, or solids in the form of dust.
- Hazardous wastes are those substances that result from a service being performed.
- Physical hazards include excessive noise, vibration, pressure, and temperature.
- Ergonomic hazards are conditions that impede normal and/or proper body position and motion.

There are many government agencies charged with ensuring safe work environments for all workers. These include the Occupational Safety and Health Administration (OSHA), Mine Safety and Health Administration (MSHA), and National Institute for Occupational Safety and Health (NIOSH). These, as well as state and local governments, have instituted regulations that must be understood and followed. Everyone in a shop has the responsibility for adhering to these regulations.

An important part of a safe work environment is the employees' knowledge of potential hazards. Right-to-know laws concerning all chemicals protect every employee in the shop. The general intent of right-to-know laws is to ensure that employers provide their employees with a safe working place as far as hazardous materials are concerned.

All employees must be trained about their rights under the legislation, the nature of the hazardous chemicals in their workplace, and the contents of the labels on the chemicals. All of the information about each chemical must be posted on material safety data sheets (MSDS) and must be accessible. The manufacturer of the chemical must give these sheets to its customers, if they are requested to do so. The sheets detail the chemical composition and precautionary information for all products that can present a health or safety hazard.

Employees must become familiar with the general uses, protective equipment, accident or spill procedures, and any other information regarding the safe handling of a particular hazardous material. This training must be given to employees annually and provided to new employees as part of their job orientation.

All hazardous material must be properly labeled, indicating what health, fire, or reactivity hazard it poses and what protective equipment is necessary when handling each chemical. The manufacturer of the hazardous materials must provide all warnings and precautionary information, which must be read and understood by the user before use. A list of all hazardous materials used in the shop must be posted for the employees to see.

Shops must maintain documentation on the hazardous chemicals in the workplace, proof of training programs, records of accidents or spill incidents, satisfaction of employee requests for specific chemical information via the MSDS, and a general right-to-know compliance procedure manual utilized within the shop.

When handling any hazardous materials or hazardous waste, make sure you follow the required procedures for handling such material. Also wear the proper safety equipment listed on the MSDS. This includes the use of approved respirator equipment.

Some of the common hazardous materials that automotive technicians use are: cleaning chemicals, fuels (gasoline and diesel), paints and thinners, battery electrolyte (acid), used engine oil, refrigerants, and engine coolant (antifreeze).

Many repair and service procedures generate what are known as hazardous wastes. Dirty solvents and cleaners are good examples of hazardous wastes. Something is classified as a hazardous waste if it is on the EPA list of known harmful materials or has one or more of the following characteristics.

- *Ignitability.* If it is a liquid with a flash point below 140°F or a solid that can spontaneously ignite.
- *Corrosivity.* If it dissolves metals and other materials or burns the skin.
- *Reactivity.* Any material that reacts violently with water or other materials or releases cyanide gas, hydrogen sulfide gas, or similar gases when exposed to low pH acid solutions. This also includes material that generates toxic mists, fumes, vapors, and flammable gases.
- *Toxicity.* Materials that leach one or more of eight heavy metals in concentrations greater than 100 times primary drinking water standard concentrations.

Complete EPA lists of hazardous wastes can be found in the Code of Federal Regulations. It should be noted that no material is considered hazardous waste until the shop is finished using it and ready to dispose of it.

The following list covers the recommended procedure for dealing with some of the common hazardous wastes. Always follow these and any other mandated procedures.

Oil Recycle oil. Set up equipment, such as a drip table or screen table with a used oil collection bucket, to collect oils dripping off parts. Place drip pans underneath vehicles that are leaking fluids onto the storage area. Do not mix other wastes with used oil, except as allowed by your recycler. Used oil generated by a shop (and/or oil received from household "do-it-yourself" generators) may be burned on site in a commercial space heater. Also, used oil may be burned for energy recovery. Contact state and local authorities to determine requirements and to obtain the necessary permits.

Oil filters Drain for at least 24 hours, crush, and recycle used oil filters.

Batteries Recycle batteries by sending them to a reclaimer or back to the distributor. Keeping shipping receipts can demonstrate that you have done the recycling. Store batteries in a watertight, acid-resistant container. Inspect batteries for cracks and leaks when they come in. Treat a dropped battery as if it were cracked. Acid residue is hazardous because it is corrosive and may contain lead and other toxic substances. Neutralize spilled acid, by using baking soda or lime, and dispose of it as hazardous material.

Metal residue from machining Collect metal filings when machining metal parts. Keep them separate and recycle if possible. Prevent metal filings from falling into a storm sewer drain.

Refrigerants Recover and/or recycle refrigerants during the servicing and disposal of motor vehicle air conditioners and refrigeration equipment. It is not allowable to knowingly vent refrigerants to the atmosphere. Recovering and/or recycling must be performed by an EPA-certified technician using certified equipment and following specified procedures.

Solvents Replace hazardous chemicals with less toxic alternatives that perform equally. For example, substitute water-based cleaning solvents for petroleum-based solvent degreasers. To reduce the amount of solvent used when cleaning parts, use a two-stage process: dirty solvent followed by fresh solvent. Hire a hazardous waste management service to clean and recycle solvents. (Some spent solvents must be disposed of as hazardous waste, unless recycled properly.) Store solvents in closed containers to prevent evaporation. Evaporation of solvents contributes to ozone depletion and smog formation. In addition, the residue from evaporation must be treated as a hazardous waste. Properly label spent solvents and store them on drip pans or in diked areas and only with compatible materials.

Containers Cap, label, cover, and properly store above ground outdoor liquid containers and small tanks within a diked area and on a paved impermeable surface to prevent spills from running into surface or ground water.

Other solids Store materials such as scrap metal, old machine parts, and worn tires under a roof or tarpaulin to protect them from the elements and to prevent the possibility of creating contaminated runoff. Consider recycling tires by retreading them.

Liquid recycling Collect and recycle coolants from radiators. Store transmission fluids, brake fluids, and solvents containing chlorinated hydrocarbons separately, and recycle or dispose of them properly.

Shop towels and rags Keep waste towels in a closed container marked "contaminated shop towels only." To reduce costs and liabilities associated with disposal of used towels, which can be classified as hazardous wastes, investigate using a laundry service that is able to treat the wastewater generated from cleaning the towels.

Waste storage Always keep hazardous waste separate, properly labeled, and sealed in the recommended containers. The storage area should be covered and may need to be fenced and locked if vandalism could be a problem. Select a licensed hazardous waste hauler after seeking recommendations and reviewing the firm's permits and authorizations.

ENGINE REPAIR TOOLS AND EQUIPMENT

Many different tools and many kinds of testing and measuring equipment are used to service engines. NATEF has identified many of these and has said an Engine Repair technician must know what they are and how and when to use them. The tools and equipment listed by NATEF are covered in the following discussion. Also included are the tools and equipment you will use while completing the job sheets. Although you need to be more than familiar with and will be using common hand tools, they are not part of this discussion. You should already know what they are and how to use and care for them.

Oil Pressure Gauge

Checking the engine's oil pressure will give you information about the condition of the oil pump, pressure regulator, and the entire lubrication system. Lower than normal oil pressures can be caused by excessive engine bearing clearances. Oil pressure is checked at the sending unit passage with an externally mounted mechanical oil pressure gauge (Figure 2). Various fittings are usually supplied with the oil pressure gauge to fit different openings in the lubrication system.

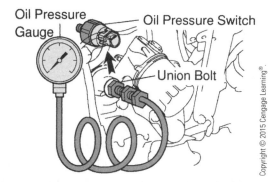

Figure 2 An oil pressure gauge can be connected to the bore for the oil pressure switch or sending unit.

Copyright © 2015 Cengage Learning®

To get accurate results from the test, make sure you follow the manufacturer's recommendations and compare your findings to specifications. Low oil pressure readings can be caused by internal component wear, pump-related problems, low oil level, contaminated oil, or low oil viscosity. An overfilled crankcase, high oil viscosity, or a faulty pressure regulator can cause high oil pressure readings.

Stethoscope

Some engine sounds can be easily heard without using a listening device, but others are impossible to hear unless they are amplified. A stethoscope is very helpful in locating engine noise because it amplifies the sound waves. It can also distinguish between normal and abnormal noise. The procedure for using a stethoscope is simple. Use the metal prod to trace the sound until it reaches its maximum intensity. Once the precise location has been discovered, the sound can be better evaluated. A sounding stick, which is nothing more than a long, hollow tube, works on the same principle, though a stethoscope gives much clearer results.

The best results, however, are obtained with an electronic listening device. With this tool you can tune into the noise. Doing this allows you to eliminate all other noises that might distract or mislead you.

Computer Memory Saver

Memory savers are an external power source used to maintain the memory circuits in electronic accessories and the engine, transmission,

and body computers when the vehicle's battery is disconnected. The saver is plugged into the vehicle's cigar lighter outlet. It can be powered by a 9- or 12-volt battery.

Hybrid Tools

A hybrid vehicle is an automobile and as such is subject to many of the same problems as a conventional vehicle. Most systems in a hybrid vehicle are diagnosed in the same way as well. However, a hybrid vehicle has unique systems that require special procedures and test equipment. It is imperative to have good information before attempting to diagnose these vehicles. Also, make sure you follow all test procedures precisely as they are given.

An important diagnostic tool is a DMM. However, this is not the same DMM used on a conventional vehicle. The meter used on hybrids and other electric-powered vehicles should be classified as a category III meter. There are basically four categories for low-voltage electrical meters, each built for specific purposes and to meet certain standards. Low-voltage, in this case, means voltages less than 1000-volts. The categories define how safe a meter is when measuring certain circuits. The standards for the various categories are defined by the American National Standards Institute (ANSI), the International Electrotechnical Commission (IEC), and the Canadian Standards Association (CSA). A CAT III meter is required for testing hybrid vehicles because of the high voltages, three-phase current, and the potential for high transient voltages. Transient voltages are voltage surges or spikes that occur in AC circuits. To be safe, you should have a CAT III-1000 V meter. A meter's voltage rating reflects its ability to withstand transient voltages. Therefore, a CAT III 1000 V meter offers much more protection than a CAT III meter rated at 600 volts.

Another important tool is an insulation resistance tester. These can check for voltage leakage through the insulation of the high-voltage cables. Obviously no leakage is desired and any leakage can cause a safety hazard as well as damage to the vehicle. Minor leakage can also cause hybrid system-related driveability problems. This meter is not one commonly used by automotive technicians, but should be for anyone who might service a damaged hybrid vehicle, such as doing body repair. This should also be a CAT III meter and may be capable of checking resistance and voltage of circuits like a DMM.

To measure insulation resistance, system voltage is selected at the meter and the probes placed at their test position. The meter will display the voltage it detects. Normally, resistance readings are taken with the circuit de-energized unless you are checking the effectiveness of the cable or wire insulation. In this case, the meter is measuring the insulation's effectiveness and not its resistance.

The probes for the meters should have safety ridges or finger positioners. These help prevent physical contact between your fingertips and the meter's test leads.

Portable Crane

To remove and install an engine, a portable crane, frequently called a cherry picker, is used. To lift an engine, attach a pulling sling or chain to the engine. Some engines have eye plates for use in lifting. If they are not available, the sling must be bolted to the engine. The sling attaching bolts must be large enough to support the engine and must thread into the block a minimum of 1-1/2 times the bolt diameter. Connect the crane to the chain. Raise the engine slightly and make sure the sling attachments are secure. Carefully lift the engine out of its compartment.

Lower the engine close to the floor so the transmission and torque converter or clutch can be removed from the engine, if necessary.

Engine Stands/Benches

After the engine has been removed, use the crane to raise the engine. Position the engine next to an engine stand. Most stands use a plate with several holes or adjustable arms. The engine must be supported by at least four bolts that fit solidly into the engine. The engine should be positioned so that its center is in the middle of the engine's stand adapter plate (Figure 3). The adapter plate can swivel in the stand. By centering the engine, it can be easily turned to the desired working positions.

Some shops have engine mounts bolted to the top of workbenches. The engine is suspended off the side of the workbench. This kind of workbench has the advantage of a good working space next to the engine, but it is not mobile and all engine work must be done at that location.

After the engine is secured to its mount, the crane and lifting chains can be removed and disassembly of the engine can begin.

Transaxle Removal and Installation Equipment

R&R of transversely mounted engines may require other tools. The engines of some FWD vehicles are removed by lifting them from the top. Others must be removed from the bottom

Figure 3 A cylinder block mounted to an engine stand.

and the procedure requires different equipment. Make sure you follow the instructions given by the manufacturer and use the appropriate tools and equipment. The required equipment varies with manufacturer and vehicle model; however, most accomplish the same thing.

To remove the engine from under the vehicle, the vehicle must be raised. A crane and/or support fixture is used to hold the engine and transaxle assembly in place while the engine is being readied for removal. Once the engine is ready, the crane is used to lower the engine onto an engine cradle. The cradle is similar to a hydraulic floor jack and is used to lower the engine to the point where it can be rolled out from under the vehicle.

Often a transverse-mounted engine is removed with the transaxle, as a unit. The transaxle can be separated from the engine once it has been lifted out of the vehicle. In this case, the drive axles must be disconnected from the transaxle before removing the unit.

Machinist's Rule

A machinist's rule is very much like an ordinary ruler. Each edge of this measuring tool is divided into increments based on a different scale. A typical machinist's rule based on the USCS system of measurement may have scales based on 1/8-, 1/16-, 1/32-, and 1/64-inch intervals. Of course, metric machinist rules are also available. Metric rules are usually divided into 0.5-mm and 1-mm increments.

Some machinist rules are based on decimal intervals. These are typically divided into 1/10-, 1/50-, and 1/1,000-inch (0.1, 0.01, and 0.001) increments. Decimal machinist rules are very helpful when measuring dimensions that are specified in decimals; they eliminate the need to convert fractions to decimals.

Micrometers

A micrometer is used to measure linear outside and inside dimensions. Both outside and inside micrometers are calibrated and read in the same manner. The major components and markings of a micrometer include the frame, anvil, spindle, locknut, sleeve, sleeve numbers, sleeve long line, thimble marks, thimble, and ratchet. Micrometers are calibrated in either inch or metric graduations and are available in a range of sizes.

To use and read a micrometer, choose the appropriate size for the object being measured. Typically they measure an inch, therefore the range covered by a micrometer of one size would be from 0 to 1 inch and another would measure 1 to 2 inches, and so on.

Open the jaws of the micrometer and slip the object between the spindle and anvil (Figure 4). While holding the object against the anvil, turn the thimble using your thumb and forefinger until the spindle contacts the object. Never clamp the micrometer tightly. Use only enough pressure on the thimble to allow the work to just fit between the anvil and spindle. To get accurate readings, you should slip the micrometer back and forth

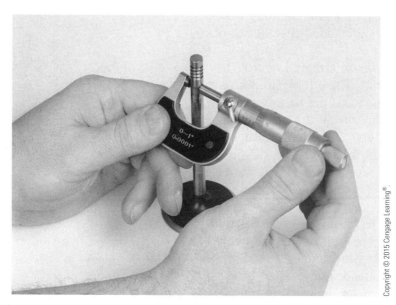

Figure 4 Using a micrometer to measure a valve stem.

over the object until you feel a very light resistance, while at the same time rocking the tool from side to side to make certain the spindle cannot be closed any further. When a satisfactory adjustment has been made, lock the micrometer and read the measurement scale.

The graduations on the sleeve each represent 0.025 inch. To read a measurement on a micrometer, begin by counting the visible lines on the sleeve and multiplying them by 0.025. The graduations on the thimble assembly define the area between the lines on the sleeve. The number indicated on the thimble is added to the measurement shown on the sleeve. This sum is the dimension of the object.

Micrometers are available that measure in 0.0001 (ten-thousandths) of an inch. Use this type of micrometer if the specifications call for this degree of accuracy.

A metric micrometer is read in the same way, except the graduations are expressed in the metric system of measurement. Each number on the sleeve represents 5 millimeters (mm) or 0.005 meter (m). Each of the ten equal spaces between each number, with index lines alternating above and below the horizontal line, represents 0.5 mm or five-tenths of an mm. Therefore, one revolution of the thimble changes the reading one space on the sleeve scale or 0.5 mm. The beveled edge of the thimble is divided into 50 equal divisions with every fifth line numbered consecutively 0, 5, 10, . . . , 45. Since one complete revolution of the thimble advances the spindle 0.5-mm, each graduation on the thimble is equal to one-hundredth of a millimeter. As with the micrometer that is graduated in inches, the separate readings are added together to obtain the total reading.

Some technicians use a digital micrometer, which is easier to read. These tools do not have the various scales; instead, the measurement is displayed and read directly off the micrometer.

Inside micrometers can be used to measure the inside diameter of a bore. To do this, place the tool inside the bore and extend the measuring surfaces until each end touches the bore's surface. If the bore is large, it might be necessary to use an extension rod to increase the micrometer's range. These extension rods come in various lengths. The inside micrometer is read in the same manner as an outside micrometer.

A depth micrometer is used to measure the distance between two parallel surfaces. The

sleeves, thimbles, and ratchet screws operate in the same way as other micrometers. Likewise, depth micrometers are read in the same way as other micrometers. If a depth micrometer is used with a gauge bar, it is important to keep both the bar and the micrometer from rocking. Any movement of either part will result in an inaccurate measurement.

Telescoping Gauge

Telescoping gauges are used for measuring bore diameters and other clearances (Figure 5). They may also be called snap gauges. They are available in sizes ranging from fractions of an inch through 6 inches. Each gauge consists of two telescoping plungers, a handle, and a lock screw. Snap gauges are normally used with an outside micrometer.

To use the telescoping gauge, insert it into the bore and loosen the lock screw. This will allow the plungers to snap against the bore. Once the plungers have expanded, tighten the lock screw. Then, remove the gauge and measure the expanse with a micrometer.

Small Hole Gauge

A small hole or ball gauge works just like a telescoping gauge. However, it is designed for small bores. After it is placed into the bore and expanded, it is removed and measured with a micrometer. Like the telescoping gauge, the small

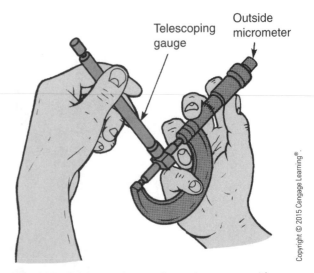

Figure 5 Measuring a telescoping gauge with an outside micrometer.

hole gauge consists of a lock, handle, and an expanding end. The end expands or retracts by turning the gauge handle.

Feeler Gauge

A feeler gauge is a thin strip of metal or plastic of known and closely controlled thickness. Several of these strips are often assembled together as a feeler gauge set that looks like a pocketknife. The desired thickness gauge can be pivoted away from the others for convenient use. A feeler gauge set usually contains strips or leaves of 0.002- to 0.010-inch thickness (in steps of 0.001 inch) and leaves of 0.012- to 0.024-inch thickness (in steps of 0.002 inch).

A feeler gauge can be used by itself to measure piston ring side clearance, piston ring end gap, connecting rod side clearance, crankshaft endplay, and other distances. It can also be used with a precision straightedge to measure main bearing bore alignment and cylinder head/block warpage.

Straightedge

A straightedge is no more than a flat bar machined to be totally flat and straight, and to be effective it must be flat and straight. Any surface that should be flat can be checked with a straightedge and feeler gauge set. The straightedge is placed across and at angles on the surface. At any low points on the surface, a feeler gauge can be placed between the straight edge and the surface. The size of the gauge that fills in the gap indicates the amount of warpage or distortion (Figure 6).

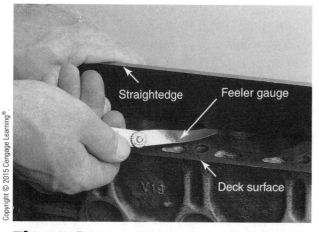

Copyright © 2015 Cengage Learning®.

Figure 6 Using a feeler gauge and straightedge to check for warpage.

Dial Indicator

The dial indicator is calibrated in 0.001-inch (one-thousandth inch) increments. Metric dial indicators are also available. Both types are used to measure movement. Common uses of the dial indicator include measuring valve lift, journal concentricity, flywheel or brake rotor runout, gear backlash, and crankshaft endplay.

To use a dial indicator, position the indicator rod against the object to be measured. Then, push the indicator toward the work until the indicator needle travels far enough around the gauge face to permit movement to be read in either direction. Zero the indicator needle on the gauge. Move the object in the direction required, while observing the needle of the gauge. Always be sure the range of the dial indicator is sufficient to allow the amount of movement required by the measuring procedure. For example, never use a 1-inch indicator on a component that will move 2 inches.

Dial Bore Gauge

Cylinder bore taper and out-of-roundness can be measured with a micrometer and telescoping gauge. However, most shops use a dial bore gauge, which typically consists of a handle, guide blocks, lock, indicator contact, and an indicator. It also comes with extensions that make it adaptable to various sized bores.

Most cylinder wear occurs at the top of the ring travel area. Pressure on the top ring is at a peak and lubrication at a minimum when the piston is at the top of its stroke. A ridge of unworn material will remain above the upper limit of ring travel. Below the ring travel area, wear is negligible because only the piston skirt contacts the cylinder wall.

A properly reconditioned cylinder must have the correct diameter, have no taper or out-of-roundness, and the surface finish must be such that the piston rings will seat to form a seal that will control oil and minimize blowby.

Taper is the difference in diameter between the bottom of the cylinder bore and the top of the bore just below the ridge. Subtracting the smaller diameter from the larger one gives the cylinder taper. On a dial bore gauge, the change in diameter will move the indicator. The highest reading on the indicator is the amount of taper in the bore.

Cylinder out-of-roundness is the difference between the cylinder's diameter when measured parallel with the crank and then perpendicular to the crank. Out-of-roundness is measured at the top of the cylinder just below the ridge. As the dial bore gauge is turned inside the bore, the indicator will show any change in the bore's diameter. The amount shown on the indicator is the amount the cylinder is out-of-round.

Ridge Reamer

After many miles of use, a ridge is formed at the top of the engine's cylinders. Because the top piston ring stops traveling before it reaches the top of the cylinder, a ridge of unworn metal is left. This ridge must be removed to remove the pistons from the block without damaging them.

This ridge is removed with a ridge reamer or ridge-removing tool. The tool must be adjusted for the bore and then inserted into it. Rotate the tool clockwise with a wrench to remove the ridge. Remove just enough metal to allow the piston assembly to slip out of the bore without causing damage to the surface of the bore or to the piston. If the ridge is too large, the top rings will hit it and possibly break the ring lands.

After the ridge removing operation, wipe all the metal cuttings out of the cylinder. Use an oily rag to wipe the cylinder. The cuttings will stick to it.

Cylinder Deglazer

The proper surface finish on a cylinder wall acts as a reservoir for oil to lubricate the piston rings and prevent piston and ring scuffing. Piston ring faces can be damaged and experience premature wear if the cylinder wall is too rough. A surface that is too smooth will not hold enough oil and won't allow the rings to seat properly. Obtaining the correct cylinder wall finish is important.

If the inspection and measurements of the cylinder wall show that surface conditions, taper, and out-of-roundness are within acceptable limits, the cylinder walls only need to be deglazed. Combustion heat, engine oil, and piston movement combine to form a thin residue on the cylinder walls that is commonly called glaze.

The common types of glaze breakers use an abrasive with about 220 or 280 grit. The glaze breaker is installed in a slow-moving electric drill or in a honing machine. Many deglazers use round stones that extend on coiled wire from the center shaft (Figure 7). This type of deglazer may

Figure 7 A resilient-based, hone-type brush, commonly called a ball hone.

also be used to lightly hone the bore. Various sizes of resilient-based hone-type brushes are available for honing and deglazing.

Cylinder Hone

You should hone a cylinder whenever there are minor problems with the bore. Honing will sand the walls to remove imperfections. A cylinder hone usually consists of two or three stones. The hone rotates at a selected speed and is moved up and down the cylinder's bore. The stones have outward pressure on them and remove some metal from the bore as they rotate within it. Honing oil flows over the stones and onto the cylinder wall to control the temperature and flush out any metallic and abrasive residue. The correct stones should be used to ensure that the finished walls have the correct surface finish. Honing stones are classified by grit size; typically the lower the grit number, the coarser the stone.

Cylinder honing machines are available in manual and automatic models. The major advantage of the automatic type is that it allows the technician to dial in the exact crosshatch angle needed.

Figure 8 A piston ring compressor.

When cylinder surfaces are badly worn or excessively scored or tapered, a boring bar is used to cut the cylinders for oversize pistons or sleeves. A boring bar leaves a pattern on the cylinder wall similar to uneven screw threads; therefore, you should hone the bore to the correct finish after it has been bored.

Ring Groove Cleaner

Before installing piston rings onto a piston, the ring grooves should be cleaned. The carbon and other debris that may be present in the back of the groove will not allow the rings to compress evenly and completely into the grooves. Piston ring grooves are best cleaned with a ring groove cleaner, which is adjustable to fit the width and depth of the groove. Make sure it is properly adjusted before using it and make sure you do not damage the piston while cleaning it.

Ring Expander

To prevent damage to the piston rings during removal and installation, a ring expander should be used. To install a piston ring, the ring must be made large enough to fit over the piston. When this is done by hand, there is a chance the ring may become distorted because it is almost impossible to evenly expand the ring with your hands. The ring expander also reduces the chances of

scratching or damaging the piston while working the ring into its proper location. The rings fit into the jaws of the expander and the handle of the tool is squeezed to expand the ring. Expand the rings only to the point where they can fit over the piston.

Ring Compressor

The use of a ring compressor is the only sensible way to install a piston with piston rings into a cylinder. The compressor wraps around the rings to make their outside diameter smaller than the inside diameter of the bore (Figure 8). With the compressor tool adjusted properly, the piston assembly can be easily pushed into the bore without damaging the bore or piston.

V-Blocks

The various shafts in an engine must be straight and free of distortion. Visually it is impossible to see any distortions unless the shaft is severely damaged. Warped or distorted shafts will cause many problems, including premature wear of the bearings they ride on. The best way to check a shaft is to place the ends of the shaft onto V-blocks. These blocks will support the shaft and allow you to rotate the shaft. Place the plunger of a dial indicator on the journals of the shaft and rotate the shaft. Any movement of the indicator's needle suggests a problem.

Cam Bearing Driver Set

The camshaft is supported by several friction-type bearings, or bushings. They are designed as one piece and are typically pressed into the camshaft bore in the cylinder head or block; however, some OHC engines use split bearings to support the camshaft. Camshaft bearings are normally replaced during engine rebuilding. The old bearings should be inspected for signs of unusual wear that may indicate an oiling or bore alignment problem.

Cam bearings are normally press fit into the block or head using a bushing driver and hammer. After the cam bearings have been installed, the oil holes in the bearings should be checked for proper alignment with those in the block or head. This will ensure correct oil supply to vital engine areas. Proper alignment can be checked by inserting a wire through the holes or by squirting oil into the holes. If the oil does not run out, the holes are misaligned. This procedure should be repeated with each bearing.

Camshaft Holding Tool

This is a special tool used to hold the camshaft in its correct position when reinstalling a timing belt or chain. The object of the tool is to not allow the camshaft to move while wrapping the belt around the pulley or sprocket. These tools are available for universal applications and for specific engines.

Valve Spring Compressor

To remove the valves from a cylinder head, the valve spring assemblies must be removed first. To do this, the valve spring must be compressed enough to remove the valve keepers, then the retainer. There are many types of spring compressors available. Some are designed to allow valve spring removal while the cylinder head is still on the engine block. Other designs are used only when the cylinder head is removed. There are also spring compressor tools designed for specific OHC and DOHC engines.

The pry bar-type compressor is used when installing valve oil seals when the cylinder head is still mounted to the block. With the cylinder's piston at TDC, shop air is fed into the cylinder to hold the valve up and prevent it falling into the cylinder. The pry bar is then used to compress the valve spring so the valve keepers can be removed.

Some OHC engines require the use of a special spring compressor. Often these special tools can be used when the cylinder head is attached to the block and when it is on a bench. These compressors bolt to the cylinder head and have a threaded plunger that fits onto the retainer. As the plunger is tightened down on the retainer, the spring compresses.

C-clamp-type valve compressors can only be used on cylinder heads after they have been removed from the block. This kind of compressor usually is a universal tool with interchangeable jaws. The spring is compressed either pneumatically or manually after the compressor is in place. One end of the clamp is positioned on the valve head and the other on the valve's retainer. After the compressor is adjusted, the compressor is activated to squeeze down on the spring. Once the spring is compressed, the valve keepers can be removed. Then the tension of the compressor is slowly released and the valve retainer and spring can be removed.

To reinstall the valve spring, insert the valve into the guide and position the valve spring inserts, valve spring, and retainer over the valve stem. With the spring compressor, compress the spring just enough to install the valve keepers into their grooves. Excess pressure may cause the retainer to damage the oil seal. Release the spring compressor and tap the valve stem with a rubber mallet to seat the keepers.

Valve Spring Tester

Before valve springs are used, they should be checked to make sure they are within specifications. Line up all of the springs on a flat surface and place a straightedge across the tops to check their freestanding height. Free height should be within 1/16 inch of manufacturer's specifications. Throw away any spring that is not within specifications.

A spring that is not square will cause side pressure on the valve stem and abnormal wear. To check squareness, set a spring upright against a carpenter's square. Rotate the spring until a gap appears between the spring and the square. Measure the gap with a feeler gauge. If the gap

Figure 9 A valve spring tester.

is more than 0.060 inch, the spring should be replaced.

Use a valve spring tester to check each valve's open and close pressure. Close pressure guarantees a tight seal. The open pressure overcomes valve train inertia and closes the valve when it should close. Spring tension must be checked at the installed spring height; therefore, if a shim is to be used, insert it under the spring on the valve spring tension gauge. Compress the spring into the installed height by pressing down on the tester's lever. The tester's gauge will reflect the pressure of the spring when compressed to the installed or valve closed height. Compare this reading to the specifications. Now compress the spring to the open height specification (Figure 9). Use the rule on the tester or a machinist rule to measure the compressed height. Read the pressure on the tester and compare this reading to the specifications. Any pressure outside the pressure range given in the specifications indicates the spring should be replaced. Low spring pressure may allow the valve to float during high-speed operation. Excessive spring tension may cause premature valve train wear. If the tension is within specifications, the spring can be installed on the valve stem.

Valve Guide Repair Tools

The amount of valve guide wear can be measured with a ball gauge and micrometer. Insert and expand the ball gauge at the top of the guide. Lock it to that diameter, remove it from the guide, and measure the ball gauge with an outside micrometer. Repeat this process with the ball gauge in the middle and the bottom of the guide. Compare your measurements to the specifications for valve guide inside diameter. Then compare your measurements against each other. Any difference in reading shows a taper or wear inside the guide.

Another way to check for excessive guide wear is with a dial indicator. The accuracy of this check is directly dependent on the amount the valve is open during the check. Some manufacturers specify this amount or provide special spacers that are installed over the valve stem to ensure the proper height. Attach the dial indicator to the cylinder head and position it so the plunger is at a right angle to the valve stem being measured. With the plunger in contact with the valve head, move the valve toward the indicator and set the dial indicator to zero. Now move the valve from the indicator. Observe the reading on the dial while doing this. The reading on the indicator is the total movement of the valve and is indicative of the guide's wear. Compare the reading to the specifications.

Valve guide wear or taper must be corrected. A certain minimum amount is needed for lubrication and thermal expansion of the valve stem. Exhaust valves require more clearance than intakes because they run hotter. Clearance should also be close enough to prevent a buildup of varnish and carbon deposits on the stems, which could cause sticking. Insufficient clearance, however, can lead to rapid stem and guide wear, scuffing, and sticking, which prevents the valve from seating fully. If the clearance is too great, oil control will be a serious problem.

Knurling is one of the fastest ways to restore the inside diameter dimensions of a worn valve guide. The procedure raises the surface of the guide ID by plowing tiny furrows through the surface of the metal. As the knurling tool cuts into the guide, metal is raised or pushed up on either side of the indentation. This effectively decreases the ID of the guide hole. A burnisher is used to make the ridges flat and produce the proper-sized hole to restore the correct guide-to-stem clearance.

Reaming is used to repair worn guides by increasing the guide hold size to take an oversize valve stem or by restoring the guide to its original diameter after installing inserts. When reaming a guide, limit the amount of metal removed and always reface the valve seat after the valve guide has been reamed.

Installing thin-wall guide liners is popular with many production engine rebuilders, as well as smaller shops. The liners are installed by first boring out the original guides to a specified amount and then pressing the liners into the guide. Some of these liners are not precut to length and the excess must be milled off before finishing.

Replacing the entire valve guide is another repair option that is possible on cylinder heads with replaceable guides. To replace integral guides, bore or drive the old guide out and drive a thin-wall replacement guide into the hole. It is important to keep the centerline of the guide concentric with the valve seat so that the rocker arm-to-valve stem contact area is not disturbed.

If the original guide can be removed and a new one inserted, press out an old valve guide. Do this by placing the properly sized driver into the guide. The shoulder on the driver must also be slightly smaller than the OD of the guide, so that it will go through the cylinder head. Then press out the guide. To install a new guide, use a press and the same driver that was used to remove the old guide. Align the new guide and press straight down, not at an angle.

Valve and Valve Seat Resurfacing Equipment

Whenever the valves have been removed from the cylinder head, the valve heads and valve seats should be resurfaced. The most critical sealing surface in the valve train is between the face of the valve and its seat in the cylinder head when the valve is closed. Leakage between these surfaces reduces the engine's compression and power, and can lead to valve burning. To insure proper seating of the valve, the seat area on the valve face and seat must be the correct width, at the correct location, and concentric with the guide. These conditions are accomplished by renewing the surface of the valve face and seat.

Valve grinding or refacing is done by machining a fresh, smooth surface on the valve faces and stem tips. Valve faces suffer from burning, pitting, and wear caused by opening and closing millions of times during the life of an engine. Valve stem tips wear because of friction from the rocker arms or actuators.

Before grinding, each valve face should be inspected for burning or distortion and each stem tip for wear. Replace any valves that are badly burned or worn. Valves can be refaced using either grinding or cutting machines. Although both can reface valves and smoothen and chamfer valve stem ends, the traditional grinding method of refacing is still the most popular.

To start the valve grinding operation, chuck the valve as close as possible to the valve head to eliminate stem flexing from wheel pressure. Set the grinding angle according to specifications or the desired angle. If an interference angle is specified, the valve grinder should be set at 1/2 to 1 degree less than the standard 30- or 45-degree face angle. Make light cuts using the full width of the grinding wheel width. Remove only enough metal to clean up the valve face.

Once the face is ground, the valve tip may also need to be ground. This is best determined by placing the valve in the cylinder head to check the stem height. The tip is ground by using a stemming stone with the valve secured and a coolant flowing over the stem to cool the valve tip and remove grit. The valve tip should be ground so that it is exactly square with the stem. Because valve tips have hardened surfaces that are up to 0.030 inch in depth, only 0.010 inch can be removed during grinding.

There is basically little difference between using a grinding tool and a cutting tool to resurface a valve face so the procedure for doing either is the same. Keep in mind that the metal removed from the valve face and valve seat increases the amount of valve stem length on the spring side of the cylinder head when the valve is seated.

Before starting valve seat work, carefully check the seats for cracks. Like valve guides, there are two types of valve seats: integral and insert. Integral seats are part of the casting. Insert seats are pressed into the head and are always used in aluminum cylinder heads.

Cracked integral seats sometimes can be repaired by installing inserts, if the crack is not too deep. Insert seats are pressed into the head and are always used in aluminum cylinder heads. Valve seats can be reconditioned or repaired by one of two methods, depending on the seat type: machining a counterbore to install an insert seat, or grinding, cutting, or machining an integral seat.

When grinding a valve seat, it is very important to select and use the correct size pilot and grind stone. The grinding wheel should be positioned and centered by inserting a properly sized pilot shaft into the valve guide (Figure 10). The seat is ground by continually raising and lowering the grinder unit on and off the seat at a rate of approximately 120 times per minute. Grinding should only continue until the seat is clean and free of defects.

After the seat is ground, valve fit is checked using machinist dye. The valve face should be coated with dye, installed in its seat, and slightly rotated. The valve is then removed and the dye pattern on the valve face and valve seat inspected. If the valve face and seat are not contacting each other evenly, or if the contact line is too high, the seat must be reground with the same stone used initially to correct the condition. If the line is too low or the width is not correct, the seat must be reground with stones of different angles.

Cutting valve seats differs from grinding only in the equipment used. Hardened valve seat

cutters replace grinding wheels for seat finishing. The basic seat cutting procedures are the same as those for grinding.

Torque Wrench

Torque is the twisting force used to turn a fastener against the friction between the threads and between the head of the fastener and the surface of the component. The fact that practically every vehicle and engine manufacturer publishes a list of torque recommendations is ample proof of the importance of using proper amounts of torque when tightening nuts or bolts. The amount of torque applied to a fastener is measured with a torque-indicating wrench or torque wrench.

A torque wrench is basically a ratchet or breaker bar with some means of displaying the amount of torque exerted on a bolt when pressure is applied to the handle. Torque wrenches are available with the various drive sizes. Sockets are inserted onto the drive and then placed over the bolt. As pressure is exerted on the bolt, the torque wrench indicates the amount of torque.

The common types of torque wrenches are available with inch-pound and foot-pound increments.

- A beam torque wrench is not highly accurate. It relies on a beam metal that points to the torque reading.
- A "click"-type torque wrench clicks when the desired torque is reached. The handle is twisted to set the desired torque reading.
- A dial torque wrench has a dial that indicates the torque exerted on the wrench. The wrench may have a light or buzzer that turns on when the desired torque is reached.
- A digital readout type displays the torque and is commonly used to measure turning effort, as well as for tightening bolts. Some designs of this type torque wrench have a light or buzzer that turns on when the desired torque is reached.

Torque Angle Gauge

Some manufacturers recommend a torque-angle method for cylinder head bolts, which requires the use of a torque angle gauge. Torque-to-yield

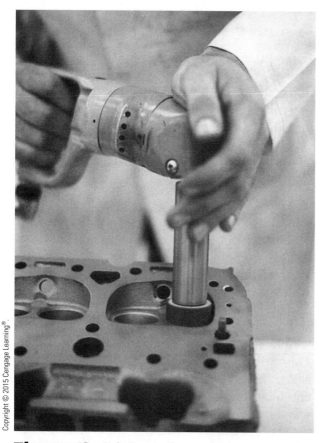

Figure 10 Refacing a valve seat with a seat grinder.

bolts must be tightened according to the manufacturer's recommendations. Typically this involves two steps: tighten the bolt to the specified torque, and then tighten the bolt an additional amount. The latter is expressed in degrees. To accurately measure the number of degrees added to the bolt, a torque angle gauge is attached to the wrench. The additional tightening will stretch the bolt, producing a very reliable clamp load that is much higher than can be achieved just by torquing.

Oil Priming Tool

Prior to starting a freshly rebuilt engine, the oil pump must be primed. There are several ways to prelubricate, or prime, an engine. One method is to drive the oil pump with an electric drill. With most engines, it is possible to make a drive that can be chucked in an electric drill motor to engage the drive on the oil pump. Insert the fabricated oil pump drive extension into the oil pump through the distributor drive hole. To control oil splash, loosely set the valve cover(s) on the engine. After running the oil pump for several minutes, remove the valve cover and see whether there is any oil flow to the rocker arms. If oil reached the cylinder head, the engine's lubrication system is full of oil and is operating properly. If no oil reached the cylinder head, there is a problem with the pump, an alignment of an oil hole in a bearing, or perhaps a plugged gallery.

Using a prelubricator, which consists of an oil reservoir attached to a continuous air supply, is the best method of prelubricating an engine without running it. When the reservoir is attached to the engine and the air pressure is turned on, the prelubricator will supply the engine's lubrication system with oil under pressure.

Cooling System Pressure Tester

A cooling system pressure tester contains a hand pump and a pressure gauge. A hose is connected from the hand pump to a special adapter that fits on the radiator filler neck. This tester is used to pressurize the cooling system and check for coolant leaks. Additional adapters are available to connect the tester to the radiator cap. With the tester connected to the radiator cap, the pressure relief action of the cap may be checked.

Coolant Hydrometer

A coolant hydrometer is used to check the amount of antifreeze in the coolant. This tester contains a pickup hose, coolant reservoir, and squeeze bulb. The pickup hose is placed in the radiator coolant. When the squeeze bulb is squeezed and released, coolant is drawn into the reservoir. As coolant enters the reservoir, a pivoted float moves upward with the coolant level. A pointer on the float indicates the freezing point of the coolant on a scale located on the reservoir housing.

Refractory Testers

For many shops, the preferred way to check coolant is with a refractometer. This tester works on the principle that light bends as it passes through a liquid. A sample of the coolant is placed in the tester. As light passes through the sample of coolant, it bends and shines on a scale in the tester. A reading is taken at the point on the scale where there is a separation of light and dark. Most refractory coolant testers can also check the electrolyte in a battery.

Measuring pH

Acids produced by bacteria and other contaminants can reduce the effectiveness of coolant. Some shops measure the pH of coolant to determine the deterioration of the coolant. The pH is measured by placing test strips or a digital pH tester into the coolant.

Combustion Leak Tester

Combustion gases can leak into an engine's coolant if there is a bad cylinder head gasket or cracked engine block or cylinder head. Exhaust from combustion can immediately destroy the inhibitors in the coolant, and set up an acid condition. The coolant can then conduct electricity, and a galvanizing reaction will begin among the various metals in the engine and cooling system. This will cause deterioration inside the radiator and other parts of the system. If the coolant enters the cylinders, the engine will run poorly and will dilute the engine's oil.

The presence of combustion gases can be checked a number of ways, however the most commonly used method involves adding a fluid to the coolant. With the engine warm and running, a sample of the coolant is drawn into an aspirator bulb and observed. Normally if a combustion leak is present, the chemical fluid in the tool will turn from blue to yellow in just a minute. If the fluid stays blue, there is no combustion leak.

Belt Tension Gauge

A belt tension gauge is used to measure drive belt tension. The belt tension gauge is installed over the belt, and the gauge indicates the amount of belt tension.

Service Information

Perhaps the most important tools you will use are service information. There is no way a technician can remember all of the procedures and specifications needed to correctly repair all vehicles. Therefore, a good technician relies on service information and other sources for this. Good information plus knowledge allows a technician to fix a problem with the least bit of frustration and at the lowest cost to the customer.

To obtain the correct engine specifications and other information, you must first identify the engine you are working on. The best source for engine identification is the VIN. The engine code can be interpreted through information given in the service information. This data also help you identify the engine through casting numbers and/or markings on the cylinder block or head.

The primary source of repair and specification information for any car, van, or truck is the manufacturer. The manufacturer publishes service manuals each year, for every vehicle built. Because of the enormous amount of information, some manufacturers publish more than one manual per year per car model. They are typically divided into sections based on the major systems of the vehicle. In the case of engines, there is a section for each engine that may be found in the vehicle. Manufacturers' manuals detail all repairs, adjustments, specifications, diagnostic procedures, and special tools required.

Since many technical changes occur on specific vehicles each year, manufacturers' service manuals need to be constantly updated. Updates are published as service bulletins (often referred to as Technical Service Bulletins or TSBs) that show the changes in specifications and repair procedures during the model year. The car manufacturer provides these bulletins to dealers and repair facilities on a regular basis.

Service information is also published by independent companies rather than by the manufacturers. However, they pay for and get most of their information from the car makers. They contain component information, diagnostic steps, repair procedures, and specifications for several car makes in one book. Information is usually condensed and is more general in nature than the manufacturer's manuals. The condensed format allows for more coverage in less space and therefore is not always specific. They may also contain several years of models as well as several car makes in one book.

Many of the larger parts manufacturers have excellent guides on the various parts they manufacture or supply. They also provide updated service bulletins on their products. Other sources for up-to-date technical information are trade magazines and trade associations.

Service information is found electronically on digital video disks (DVDs) and on the Internet. A single compact disk can hold a quarter million pages of text, eliminating the need for a huge library to contain all of the printed manuals. Using electronics to find information is also easier and quicker. All a technician needs to do is enter vehicle information and then move to the appropriate part or system. The appropriate information will then appear on the computer's screen. Online data can be updated instantly and requires no space for physical storage. These systems are easy to use and the information is quickly accessed and displayed. The computer's keyword, and mouse are used to make selections from the screen's menu. Once the information is retrieved, a technician can read it off the screen or print it out and take it to the service bay.

JOB SHEETS

REQUIRED SUPPLEMENTAL TASKS (RST)

JOB SHEET 1

Shop Safety Survey

Name _____ Station _____ Date _____

NATEF Correlation

This Job Sheet addresses the following **RST** tasks:

Shop and Personal Safety:

1. Identify general shop safety rules and procedures.

8. Identify the location and use of eyewash stations.

10. Comply with the required use of safety glasses, ear protection, gloves, and shoes during lab/shop activities

11. Identify and wear appropriate clothing for lab/shop activities

12. Secure hair and jewelry for lab/shop activities.

Objective

As a professional technician, safety should be one of your first concerns. This job sheet will increase your awareness of shop safety rules and safety equipment. As you survey your shop area and answer the following questions, you will learn how to evaluate the safeness of your workplace.

Materials

Copy of the shop rules from the instructor

PROCEDURE

Your instructor will review your progress throughout this worksheet and should sign off on the sheet when you complete it.

1. Have your instructor provide you with a copy of the shop rules and procedures.

   Have you read and understood the shop rules? ☐ Yes ☐ No

29

2. Before you begin to evaluate your work area, evaluate yourself. Are you dressed to work safely? ☐ Yes ☐ No

   If no, what is wrong? _____

3. Are your safety glasses OSHA approved? ☐ Yes ☐ No

   Do they have side protection shields? ☐ Yes ☐ No

4. Look around the shop and note any area that poses a potential safety hazard.

   _____

   _____

   All true hazards should be brought to the attention of the instructor immediately.

5. What is the air line pressure in the shop? _____ psi

   What should it be? _____ psi

6. Where is the first-aid kit(s) kept in the work area?

   _____

   _____

7. Ask the instructor to show the location of, and demonstrate the use of, the eyewash station. Where is it and when should it be used?

   _____

   _____

8. What is the shop's procedure for dealing with an accident?

   _____

   _____

   _____

9. Explain how to secure hair and jewelry while working in the shop.

   _____

10. List the phone numbers that should be called in case of an emergency.

   _____

Problems Encountered

_____

_____

_____

Instructor's Comments

_____

_____

_____

JOB SHEET 2

Working in a Safe Shop Environment

Name _____ Station _____ Date _____

NATEF Correlation

This Job Sheet addresses the following **RST** tasks:

Shop and Personal Safety:

2. Utilize safe procedures for handling of tools and equipment.

3. Identify and use proper placement of floor jacks and jack stands.

4. Identify and use proper procedures for safe lift operation.

5. Utilize proper ventilation procedures for working within the lab/shop area.

6. Identify marked safety areas.

9. Identify the location of the posted evacuation routes.

Objective

This job sheet will help you work safely in the shop. Two of the basic tools a technician uses are lifts and the floor jack.

This job sheet also covers the technicians' environment.

Materials

Vehicle for hoist and jack stand demonstration
Service information

Describe the vehicle being worked on:

Year _____ Make _____ Model _____

VIN _____ Engine type and size _____

PROCEDURE

1. Are there safety areas marked around grinders and other machinery?   ☐ Yes   ☐ No

2. Are the shop emergency escape routes clearly marked?   ☐ Yes   ☐ No

3. Have your instructor demonstrate the exhaust gas ventilation system in the shop. Explain the importance of the ventilation system.

   _____

   _____

4. What types of Lifts are used in the shop?

_____

_____

5. Find the location of the correct lifting points for the vehicle supplied by the instructor. On the rear of this sheet, draw a simple figure showing where these lift points are.

6. Ask your instructor to demonstrate the proper use of the lift.

   Summarize the proper use of the lift.

   _____

   _____

   _____

   _____

7. Demonstrate the proper use of jack stands with the help of your instructor.

   Summarize the proper use of jack stands.

   _____

   _____

   _____

Problems Encountered

_____

_____

_____

Instructor's Comments

_____

_____

_____

JOB SHEET 3

Fire Extinguisher Care and Use

Name _____ Station _____ Date _____

NATEF Correlation

This Job Sheet addresses the following **RST** task:

Shop and Personal Safety:

7. Identify the location and the types of fire extinguishers and other fire safety equipment; demonstrate knowledge of the procedures for using fire extinguishers and other fire safety equipment.

Objective

Upon completion of this job sheet, you will be able to demonstrate knowledge of the procedures for using fire extinguishers and other fire safety equipment, and identify the location of fire extinguishers in the shop.

NOTE: *Never fight a fire that is out of control or too large. Call the fire department immediately!*

1. Identify the location of the fire extinguishers in the shop.

   _____

   _____

2. Have the fire extinguishers been inspected recently? (Look for a dated tag.)

   _____

   _____

3. What types of fires are each of the shop's fire extinguishers rated to fight?

   _____

   _____

4. What types of fires should not be used with the shop's extinguishers?

   _____

   _____

5. One way to remember the operation of a fire extinguisher is to remember the term PASS. Describe the meaning of PASS on the following lines.

   a. P

   _____

   b. A

   _____

c. S

_____

d. S

_____

Problems Encountered

_____

_____

_____

Instructor's Comments

_____

_____

_____

JOB SHEET 4

Working Safely Around Air Bags

Name _____ Station _____ Date _____

NATEF Correlation

This Job Sheet addresses the following **RST** task:

Shop and Personal Safety:

13. Demonstrate awareness of the safety aspects of supplemental restraint systems (SRS), electronic brake control systems, and hybrid vehicle high-voltage circuits.

Objective

Upon completion of this job sheet, you should be able to work safely around and with air bag systems.

Tools and Materials

A vehicle(s) with air bag

Safety glasses, goggles

Service information appropriate to vehicle(s) used

Describe the vehicle being worked on:

Year _____ Make _____ Model _____

VIN _____ Engine type and size _____

PROCEDURE

1. Locate the information about the air bag system in the service information. How are the critical parts of the system identified in the vehicle?

   _____

   _____

2. List the main components of the air bag system and describe their location.

   _____

   _____

3. There are some very important guidelines to follow when working with and around air bag systems. Look through the service information to find the answers to the questions and fill in the blanks with the correct words.

   a. Wait at least _____ minutes after disconnecting the battery before beginning any service. The reserve _____ module is capable of storing enough energy to deploy the air bag for up to _____ minutes after battery voltage is lost.

b. Never carry an air bag module by its _____ or _____, and, when carrying it, always face the trim and air bag _____ from your body. When placing a module on a bench, always face the trim and air bag _____.

c. Deployed air bags may have a powdery residue on them. _____ is produced by the deployment reaction and is converted to _____ when it comes in contact with the moisture in the atmosphere. Although it is unlikely that harmful chemicals will still be on the bag, it is wise to wear _____ and _____ when handling a deployed air bag. Immediately wash your hands after handling a deployed air bag.

d. A live air bag must be _____ before it is disposed. A deployed air bag should be disposed of in a manner consistent with the _____ and manufacturer's procedures.

e. Never use a battery- or AC-powered _____, _____, or any other type of test equipment in the system unless the manufacturer specifically says to. Never probe with a _____ for voltage.

4. Explain how an air bag sensor should be handled before it is installed on the vehicle.

Problems Encountered

_____

_____

_____

Instructor's Response

_____

_____

_____

JOB SHEET 5

High-Voltage Hazards in Today's Vehicles

Name _____ Station _____ Date _____

NATEF Correlation

This Job Sheet addresses the following **RST** task:

Shop and Personal Safety:

14. Demonstrate awareness of the safety aspects of high-voltage circuits (such as high intensity discharge (HID) lamps, ignition systems, injection systems, etc.).

Objective

Upon completion of this job sheet, you will be able to describe some of the necessary precautions to take in performing work around high-voltage hazards such as HID headlamps and ignition systems.

HID Headlamp Precautions

Describe the vehicle being worked on:

Year _____ Make _____ Model _____

VIN _____ Engine type and size _____

Materials needed

Service information for HID headlamps

Describe general operating condition:

1. List three precautions a technician should observe when working with HID headlamp systems.

   _____

   _____

   _____

2. How many volts are necessary to initiate and maintain the arc inside the bulb of this HID system?

   _____

   _____

3. A technician must never probe with a test lamp between the HID ballast and the bulb. Explain why?

   _____

High-Voltage Ignition System Precautions

4. High-voltage ignition systems can cause serious injury, especially for those who have heart problems. Name at least three precautions to take when working around high-voltage ignition systems.

_____

_____

_____

_____

Problems Encountered

_____

_____

_____

Instructor's Comments

_____

_____

_____

JOB SHEET 6

Hybrid High-Voltage and Brake System Pressure Hazards

Name _____ Station _____ Date _____

NATEF Correlation

This Job Sheet addresses the following **RST** task:

Shop and Personal Safety:

13. Demonstrate awareness of the safety aspects of supplemental restraint systems (SRS), electronic brake control systems, and hybrid vehicle high-voltage circuits.

Objective

Upon completion of this job sheet, you will be able to describe some of the necessary precautions to take while performing work around high-voltage systems such as that used in hybrid vehicles, as well as the high-pressure systems of some braking control systems.

Tools and Materials

Appropriate service information

AUTHOR'S NOTE: *According to the vehicles' manufacturers, a technician should not work on hybrid vehicles without having the specific training, which is beyond the scope of this job sheet. This job sheet assumes that the student will be accessing information only, not actually working on live high-voltage vehicles.*

Protective Gear

Goggles or safety glasses with side shields

High-voltage gloves with properly inspected liners

Orange traffic cones to warn others in the shop of a high-voltage hazard

Describe the vehicle being worked on:

Year _____ Make _____ Model _____

VIN _____ Engine type and size _____

Describe general operating condition:

_____

PROCEDURE

High Pressure Braking Systems:

NOTE: *Many vehicles have a high pressure accumulator or a high pressure pump in their braking systems. Opening one of these systems can be hazardous due to the high pressures involved.*

1. Research the vehicle you have been assigned and describe the procedure that must be followed BEFORE opening the hydraulic braking system.

   _____

   _____

   _____

Hybrid Vehicles:

1. Special gloves with liners are to be inspected before each use when working on a hybrid vehicle.

   A. How should the integrity of the gloves be checked?

   _____

   _____

   B. How often should the gloves be tested (at a lab) for recertification?

   _____

   C. What precautions must be taken when storing the gloves?

   _____

   _____

   _____

   D. When must the gloves be worn?

   _____

   _____

   _____

2. What color are the high-voltage cables on hybrid vehicles?

   _____

3. What must be done BEFORE disconnecting the main voltage supply cable?

   _____

   _____

   _____

4. Describe the safety basis of the "one hand rule."

   _____

   _____

   _____

5. Explain the procedure to disable high voltage on the vehicle you selected.

   _____

   _____

   _____

6. Explain the procedure to test for high voltage to ensure that the main voltage is disconnected.

   _____

   _____

Problems Encountered

_____

_____

_____

Instructor's Comments

_____

_____

_____

JOB SHEET 7

Material Data Safety Sheet Usage

Name _____ Station _____ Date _____

NATEF Correlation

This Job Sheet addresses the following **RST** task:

Shop and Personal Safety:

12. Locate and demonstrate knowledge of material safety data sheets (MSDS).

Objective

On completion of this job sheet, the student will be able to locate the MSDS folder and describe the use of an MSDS sheet on the job site.

Materials

Selection of chemicals from the shop
MSDS sheets

1. Locate the MSDS folder in the shop. It should be in a prominent location. Did you have any problems finding the folder?

   _____

   _____

2. Pick a common chemical from your tool room such as brake cleaner. Locate the chemical in the MSDS folder. What chemical did you choose?

   _____

   _____

3. What is the flash point of the chemical? _____

4. Briefly describe why the flash point is important.

   _____

   _____

5. What is the first aid if the chemical is ingested?

   _____

   _____

6. Can this chemical be absorbed through the skin?

   _____

   _____

7. What are the signs of exposure to the chemical you selected?

_____

_____

8. What are the primary health hazards of the chemical?

_____

_____

9. What is the first aid procedure for exposure to this chemical?

_____

_____

10. What are the recommendations for protective clothing?

_____

_____

Problems Encountered

_____

_____

_____

Instructor's Comments

_____

_____

_____

JOB SHEET 8

Measuring Tools and Equipment Use

Name _____ Station _____ Date _____

NATEF Correlation

This Job Sheet addresses the following **RST** tasks:

Tools and Equipment:

1. Identify tools and their usage in automotive applications.

2. Identify standard and metric designations.

3. Demonstrate safe handling and use of appropriate tools.

4. Demonstrate proper cleaning, storage, and maintenance of tools and equipment.

5. Demonstrate proper use of precision measuring tools (i.e., micrometer, dial-indicator, dial caliper).

Objective

Upon completion of this job sheet, you will be able to make measurements using micrometers, dial indicators, pressure gauges, and other measuring tools. You will also be able to demonstrate the safe handling and use of appropriate tools, and demonstrate their proper use.

Tools and Materials

Items to measure selected by the instructor

Precision measuring tools: micrometer (digital or manual), dial caliper, vacuum or pressure gauge (selected by the instructor)

PROCEDURE

Have your instructor demonstrate the measuring tools that are available to you in your shop. The tools should include both standard and metric tools.

1. Describe the measuring tools you will be using.

   _____

   _____

2. Describe the items that you will be measuring.

   _____

   _____

3. Describe any special handling or safety procedures that are necessary when using the tools.

   _____

   _____

4. Describe any special cleaning or storage that these tools might require.

_____

_____

5. Describe the metric unit of measurement of the tools, such as millimeters, centimeters, kilopascals, etc.

_____

_____

6. Describe the standard unit of measure of the tools, such as inches, pounds per square inch, etc.

_____

_____

7. Measure the components, list them, and record your measurements in the following table.

Item measured	Measurement taken	Unit of measure

8. Clean and store the tools. Describe the process next.

Problems Encountered

_____

_____

_____

Instructor's Comments

_____

_____

_____

JOB SHEET 9

Preparing the Vehicle for Service and Customer

Name _____ Station _____ Date _____

NATEF Correlation

This Job Sheet addresses the following **RST** tasks:

Preparing Vehicle for Service:

1. Identify information needed and the service requested on a repair order.

2. Identify purpose and demonstrate proper use of fender covers, mats.

3. Demonstrate use of the three C's (concern, cause, and correction).

4. Review vehicle service history.

5. Complete work order to include customer information, vehicle identifying information, customer concern, related service history, cause, and correction.

Preparing Vehicle for Customer:

1. Ensure vehicle is prepared to return to customer per school/company policy (floor mats, steering wheel cover, etc.).

Objective

Upon completion of this job sheet, you will be able to prepare a service work order based on customer input, vehicle information, and service history. The student will also be able to describe the appropriate steps to take to protect the vehicle and delivering the vehicle to the customer after the repair.

Tools and Materials

An assigned vehicle or the vehicle of your choice

Service work order or computer-based shop management package

Parts and labor guide

Work Order Source: Describe the system used to complete the work order. If a paper repair order is being used, describe the source.

_____

_____

PROCEDURE

1. Prepare the shop management software for entering a new work order or obtain a blank paper work order. Describe the type of repair order you are going to use.

   _____

   _____

2. Enter customer information, including name, address, and phone numbers onto the work order.

   Task Completed ☐

3. Locate and record the vehicle's VIN. Where did you find the VIN?

   _____

4. Enter the necessary vehicle information, including year, make, model, engine type and size, transmission type, license number, and odometer reading.

   Task Completed ☐

5. Does the VIN verify that the information about the vehicle is correct?

   _____

6. Normally, you would interview the customer to identify his or her concerns. However, to complete this job sheet, assume the only concern is that the customer wishes to have the front brake pads replaced. Also, assume no additional work is required to do this. Add this service to the work order.

   Task Completed ☐

7. Prepare the vehicle for entering the service department. Add floor mats, seat covers, and steering wheel covers to the vehicle.

   Task Completed ☐

8. The history of service to the vehicle can often help diagnose problems, as well as indicate possible premature part failure. Gathering this information from the customer can provide some of the data needed. For this job sheet, assume the vehicle has not had a similar problem and was not recently involved in a collision. Service history is further obtained by searching files for previous service. Often this search is done by customer name, VIN, and license number. Check the files for any related service work.

   Task Completed ☐

9. Search for technical service bulletins on this vehicle that may relate to the customer's concern. Did you find any? _____ If so, record the reference numbers here.

   _____

   _____

10. Based on the customer's concern, service history, TSBs, and your knowledge, what is the likely cause of this concern?

    _____

    _____

11. Add this information to the work order.   Task Completed ☐

12. Prepare to make a repair cost estimate for the customer. Identify all parts that may need to be replaced to correct the concern. List these here.

_____

_____

13. Describe the task(s) that will be necessary to replace the part.

_____

_____

14. Using the parts and labor guide, locate the cost of the parts that will be replaced and enter the cost of each item onto the work order at the appropriate place for creating an estimate. If the valve or cam cover is leaking, what part will need to be replaced?

_____

15. Now, locate the flat rate time for work required to correct the concern. List each task with its flat rate time.

_____

_____

16. Multiply the time for each task by the shop's hourly rate and add the cost of each item to the work order at the appropriate place for creating an estimate. Ask your instructor which shop labor rate to use and record it here.

_____

17. Many shops have a standard amount they charge each customer for   Task Completed ☐
    shop supplies and waste disposal. For this job sheet, use an amount of
    ten dollars for shop supplies.

18. Add the total costs and insert the sum as the subtotal of the estimate.   Task Completed ☐

19. Taxes must be included in the estimate. What is the sales tax rate and does it apply to both parts and labor, or just one of these?

_____

_____

20. Enter the appropriate amount of taxes to the estimate, then add this to the   Task Completed ☐
    subtotal. The end result is the estimate to give the customer.

21. By law, how accurate must your estimate be?

_____

_____

22. Generally speaking, the work order is complete and is ready for the customer's signature. However, some businesses require additional information; make sure you add that information to the work order. On the work order, there is a legal statement that defines what the customer is agreeing to. Briefly describe the contents of that statement.

_____

23. Now that the vehicle is completed and it is ready to be returned to the customer, what are the appropriate steps to take to deliver the vehicle to the customer? What would you do to make the delivery special? What should not happen when the vehicle is delivered to the customer?

_____

_____

Problems Encountered

_____

_____

_____

Instructor's Comments

_____

_____

_____

JOB SHEETS

ENGINE REPAIR JOB SHEET 10

Filling Out a Work Order

Name _____ Station _____ Date _____

NATEF Correlation

This Job Sheet addresses the following **MLR** task:

A.1. Research applicable vehicle and service information, vehicle service history, service precautions, and technical service bulletins.

This Job Sheet addresses the following **AST/MAST** tasks:

A.1. Complete work order to include customer information, vehicle identifying information, customer concern, related service history, cause, and correction.

A.2. Research applicable vehicle and service information, vehicle service history, service precautions, and technical service bulletins.

Objective

Upon completion of this job sheet, you will be able to prepare a service work order based on customer input, vehicle information, and service history.

Tools and Materials

An assigned vehicle or the vehicle of your choice

Service work order or computer-based shop management package

Parts and labor guide

Work Order Source: Describe the system used to complete the work order. If a paper repair order is being used, describe the source.

_____

_____

PROCEDURE

1. Prepare the shop management software for entering a new work order or obtain a blank paper work order. Which will you be using?

   _____

   _____

2. Enter customer information, including name, address, and phone numbers onto the work order.   Task Completed ☐

3. Locate and record the vehicle's VIN. What is the VIN?

   _____

4. Enter the necessary vehicle information, including year, make, model, engine type and size, transmission type, license number, and odometer reading.   Task Completed ☐

5. Does the VIN verify that the information about the vehicle is correct?

   _____

6. Normally, you would interview the customer to identify his or her concerns. However to complete this job sheet, assume the only concern is that the valve (cam) cover is leaking oil. This concern should be added to the work order.   Task Completed ☐

7. The history of service to the vehicle can often help diagnose problems as well as indicate possible premature part failure. Gathering this information from the customer can provide some of this information. For this job sheet, assume the vehicle has not had a similar problem and was not recently involved in a collision. Service history is further obtained by searching files for previous service. Often, this search is done by customer name, VIN, and license number. Check the files for any related service work.   Task Completed ☐

8. Search for technical service bulletins for this vehicle that may relate to the customer's concern. Were there any? If so, will they help with the repair?

   _____

   _____

9. Based on the customer's concern, service history, TSBs, and your knowledge, what is the likely cause of this concern?

   _____

   _____

10. Enter this information onto the work order.   Task Completed ☐

11. Prepare to make a repair cost estimate for the customer. Identify all parts that may need to be replaced to correct the concern. List these here.

    _____

    _____

Describe the task(s) necessary to replace the part.

_____

_____

12. What are the necessary precautions listed in performing the repair of the oil leak?

_____

_____

_____

13. Using the parts and labor guide, locate the cost of the parts that will be replaced and enter the cost of each item onto the work order at the appropriate place for creating an estimate. What is the total part cost?

_____

14. Now, locate the flat rate time for work required to correct the concern. List each task with its flat rate time.

_____

_____

15. Multiply the time for each task by the shop's hourly rate and enter the cost of each item onto the work order at the appropriate place for creating an estimate. What is the shop's labor rate?

_____

16. Many shops have a standard amount they charge each customer for shop supplies and waste disposal. For this job sheet, use an amount of ten dollars for shop supplies.

Task Completed ☐

17. Add the total costs and insert the sum as the subtotal of the estimate. What is the subtotal due?

_____

18. Taxes must be included in the estimate. What is the sales tax rate and does it apply to both parts and labor, or just one of these?

_____

_____

19. Enter the appropriate amount of taxes to the estimate, and then add this to the subtotal. The end result is the estimate to give the customer. What is the final estimated cost?

_____

20. By law, how accurate must your estimate be?

_____

_____

21. Generally speaking, the work order is complete and is ready for the customer's signature. However, some businesses require additional information. Make sure you add that information to the work order. On the work order, there is a legal statement that defines what the customer is agreeing to. Briefly describe the contents of that statement.

_____

_____

Problems Encountered

_____

_____

_____

Instructor's Comments

_____

_____

_____

ENGINE REPAIR JOB SHEET 11

Verifying the Operation of the Instrument Panel Warning Indicators

Name _____ Station _____ Date _____

NATEF Correlation

This Job Sheet addresses the following **MLR** task:

A.2. Verify operation of the instrument panel engine warning indicators.

This Job Sheet addresses the following **AST/MAST** task:

A.3. Verify operation of the instrument panel engine warning indicators.

Objective

Upon completion of this job sheet, you will be able to verify the operation of the instrument panel warning indicators.

Tools and Materials

Service information

Scan tool

Protective Clothing

Goggles or safety glasses with side shields

Describe the vehicle being worked on:

Year _____ Make _____ Model _____

VIN _____ Engine type and size _____

PROCEDURE

1. Using service information, determine the proper operation of the warning indicators during the "bulb test" sequence. List the time that each bulb should come on during the bulb test.

   Temperature warning _____

   Oil pressure _____

   Low coolant level _____

2. Did all the indicators come on (as equipped)? Describe what happened.

   _____

   _____

3. Most late-model gauges operate from information sent from the ECM/PCM to the instrument panel. Earlier vehicles operated from sending units wired directly to the instrument panel gauges. Using service information, briefly describe how your assigned vehicle's gauges operate.

_____

_____

_____

_____

4. If your vehicle uses the vehicle network to illuminate the warning lamps, check for any DTCs for the instrument panel warning system. What did you find?

_____

_____

_____

5. Describe the operation of the temperature warning lamp. Include the temperature that turns on the warning lamp.

_____

_____

_____

_____

6. Describe the process to verify the correct operation of the oil pressure warning lamp.

_____

_____

_____

7. Describe the operation of the low-coolant warning lamp (if equipped).

_____

_____

_____

8. If the vehicle network is used for the engine warning lamps, can you turn these lamps on with the scan tool?

_____

9. If you can turn on the warning lamp with the scan tool, how would this help you diagnose a warning lamp circuit?

_____

_____

10. Describe the diagnostic process if the oil pressure warning indicator lamp or warning stays on all the time.

_____

_____

_____

11. If your vehicle has a driver information center, describe the types of messages or information displayed concerning engine operation.

_____

_____

_____

Problems Encountered

_____

_____

_____

Instructor's Comments

_____

_____

_____

ENGINE REPAIR JOB SHEET 12

Inspecting an Engine for Leaks

Name _____ Station _____ Date _____

NATEF Correlation

This Job Sheet addresses the following **MLR** task:

A.3. Inspect engine assembly for fuel, oil, coolant, and other leaks; determine necessary action.

This Job Sheet addresses the following **AST/MAST** task:

A.4. Inspect engine assembly for fuel, oil, coolant, and other leaks; determine necessary action.

Objective

Upon completion of this job sheet, you will be able to inspect an engine assembly for fuel, oil, coolant, and other leaks.

Tools and Materials

A road-worthy car

Protective Clothing

Goggles or safety glasses with side shields

Describe the vehicle being worked on:

Year _____ Make _____ Model _____

VIN _____ Engine type and size _____

PROCEDURE

Engine Leak Diagnosis

1. Park the vehicle in a place where the floor is fairly clean and dry. Shut the engine off and allow it to cool. Once the engine is cool, look on the shop floor for any evidence of leaks. Remember, engine leaks can cause safety, drivability, and durability concerns. Describe the location and color of the fluid that leaked onto the floor.

   _____

   _____

   _____

   WARNING: *Gasoline fumes are extremely dangerous! If ignited, they will cause a very serious explosion and fire, resulting in personal injury and property damage. If you suspect that the leak is a fuel leak, immediately inform your instructor of the problem and take all precautions to avoid igniting the fuel.*

2. Engine fuel leaks are expensive and dangerous, and they should be corrected immediately when they are detected. If a gasoline odor occurs inside or near a vehicle, it should be inspected for fuel leaks immediately. Inspect the following parts of the vehicle to locate the source of the fuel leak. Describe what you found at each of these components:

a. Fuel tank:

_____

b. Fuel tank filler cap:

_____

c. Fuel lines and filter:

_____

d. EVAP system lines:

_____

e. Pressure regulator, fuel rail, and injectors:

_____

3. Based on the preceding inspection, what do you recommend to correct the fuel leak?

_____

_____

4. Engine oil leaks may cause an engine to run out of oil, resulting in serious engine damage. Sometimes the leak goes unnoticed, and yet the engine's oil needs to be added to more often than normal. A careful inspection may help you locate the source of the leak; however, if it is difficult to locate the exact cause of an oil leak, clean the engine first. Carefully look at the following engine parts and describe what you see. If oil is on or around any of these parts, look carefully at the component directly above the leak. Oil runs down the engine surfaces. Also, if an oil leak is found, check the PCV system for blockage and proper operation. A bad PCV system will allow crankcase pressure to build, and this may be the ultimate cause of the oil leak.

a. Rear main bearing seal:

_____

b. Expansion plug in rear camshaft bearing:

_____

c. Rear oil gallery plug:

_____

d. Oil pan:

_____

e. Oil filter:

_____

f. Rocker arm covers:

_____

g. Intake manifold front and rear gaskets (V-type engines):

_____

h. Timing gear cover or seal:

_____

i. Front main bearing:

_____

j. Oil pressure sending unit:

_____

k. Engine casting porosity:

_____

l. Oil cooler or lines (where applicable):

_____

5. Based on the preceding inspection, what do you recommend should be done to the vehicle to correct the oil leak?

_____

6. If an engine uses excessive oil and there is no evidence of leaks, the oil may be burning in the combustion chambers. If excessive amounts of oil are being burned in the combustion chambers, the exhaust will contain blue smoke, and the spark plugs may be fouled with oil. Excessive oil burning in the combustion chambers may be caused by worn rings and cylinders or worn valve guides and valve seals. Remove the spark plugs and describe the condition of each.

#1 _____

#2 _____

#3 _____

#4 _____

#5 _____

#6 _____

#7 _____

#8 _____

7. When an engine coolant leak causes the coolant level to be low, the engine quickly overheats and severe engine damage may occur. When you suspect a coolant leak, you should check the following components. Check them now and describe what you find.

   a. Upper radiator hose:

   _____

   b. Lower radiator hose:

   _____

   c. Heater hoses:

   _____

   d. By-pass hose:

   _____

   e. Water pump:

   _____

   f. Engine expansion plugs or block heater:

   _____

   g. Radiator:

   _____

   h. Thermostat housing:

   _____

   i. Heater core:

   _____

8. Based on the preceding inspection, what do you recommend to correct the oil leak?

   _____

   _____

9. Some coolant leaks are internal to the engine and will not be visible. A whitish exhaust may be indicative of this sort of coolant leak if a leaking head gasket or cracked engine components cause it. Take a look at the exhaust while the engine is idling and describe it.

   _____

   _____

10. A cooling system pressure tester is commonly used to locate leaks. The use of this tester is covered in another job sheet in this manual. What page is the procedure found on?

   _____

Problems Encountered

_____

_____

_____

Instructor's Comments

_____

_____

_____

ENGINE REPAIR JOB SHEET 13

Installing Engine Covers

Name _____ Station _____ Date _____

NATEF Correlation

This Job Sheet addresses the following **MLR** task:

A.4. Install engine covers using gaskets, seals and sealers as required.

This Job Sheet addresses the following **AST/MAST** task:

A.5. Install engine covers using gaskets, seals and sealers as required.

Objective

Upon completion of this job sheet, you will be able to install engine covers using the correct gaskets and sealants.

Tools and Materials

Straightedge

Flashlight

Drivers

Block of wood

Protective Clothing

Goggles or safety glasses with side shields

Describe the vehicle being worked on:

Year _____ Make _____ Model _____

VIN _____ Engine type and size _____

PROCEDURE

WARNING: *Make sure every sealant you use is oxygen-sensor safe.*

1. Identify the engine cover that is leaking and clean the area around the cover. What gaskets will you replace?

_____

_____

2. Remove or move aside all parts that may interfere with the removal of the part to be repaired. Make sure to reinstall them when the repair is completed.

Task Completed ☐

3. Loosen the cover bolts in the order described in the service information. Normally, the loosening sequence is the opposite of the tightening order. Briefly describe the order in which the bolts should be loosened.

_____

_____

4. Visually inspect all bolts as you remove them. Threads must be clean and undamaged. Discard all bolts that are not acceptable. Describe the condition of the bolts:

_____

_____

5. Gather the new gaskets and sealants for the engine. Protect the new gaskets by keeping them in their package until installation. What sealants are recommended for this job?   Task Completed ☐

6. Make sure all surfaces are free of dirt, oil deposits, rust, old sealer, and gasket material.   Task Completed ☐

Valve (Cam) Cover Gasket

1. Before installing the valve cover, make sure the cover's sealing flange is flat. Describe how to check the cover for flatness.

_____

_____

2. If the cover is cast aluminum and is distorted, what should be done?

_____

_____

3. Compare the new gasket with the old one. Make sure it matches. Also, make sure you have any seals that may also need to be replaced.   Task Completed ☐

4. Mount the gasket on the cover and align it in position. If the gasket has mounting tabs, use them in tandem with the contact adhesive.   Task Completed ☐

5. Install the bolts and torque them to specifications and in the recommended order. The specifications are:

_____

_____

6. Check and correct the oil level of the engine.   Task Completed ☐

Timing Chain Cover

> **NOTE:** *It will be necessary to remove the drive belts and the vibration damper (harmonic balancer) from the crankshaft before removing the timing cover. The procedure for that task is covered in a later Job Sheet.*

1. After removing the timing cover, remove the old gaskets and seals from the timing cover and engine block.

   Task Completed ☐

2. Check the condition of the cover and its flanges. Describe its condition.

   _____

   _____

3. Check the condition of the balancer shaft or crankshaft pulley hub. Make sure the surface is smooth. If the surface is not smooth, the seal will not be able to seal. What can be done if the pulley hub has a groove in it?

   _____

   _____

4. Compare the new gaskets and seals with the old ones.

   Task Completed ☐

5. Install a new crankshaft seal into the cover using a press, seal driver, or hammer, and a clean block of wood. When installing the seal, be sure to support the cover underneath to prevent damage.

   Task Completed ☐

6. Position the gasket on the cover. Make sure the bolt holes are aligned. Check the service information to see what type of adhesive or sealant you should use and where. Record the manufacturer's recommendations.

   _____

   _____

7. Finally, mount the timing cover and hand tighten the mounting bolts.

   Task Completed ☐

8. Tighten all of the bolts to specifications. What are the specifications?

   _____

   _____

9. Install the vibration damper (harmonic balancer) by using a special installation tool. In most cases, the damper is installed until it bottoms out against the oil slinger and the timing sprocket. Check the service manual to see how the damper should be installed. Describe the procedure here.

   _____

   _____

   _____

10. Some vibration dampers are held to the crankshaft by a retaining bolt. Be sure to install the large washer behind the retaining bolt on these engines. Tighten this bolt to specifications. The specifications are:

    _____

    _____

11. Install all parts that may have been removed or disconnected to gain access to the cover.   Task Completed ☐

Problems Encountered

_____

_____

_____

Instructor's Comments

_____

_____

_____

ENGINE REPAIR JOB SHEET 14

Inspect and Replace Timing Belts

Name _____ Station _____ Date _____

NATEF Correlation

This Job Sheet addresses the following **MLR** task:

A.5. Remove and replace timing belt; verify correct camshaft timing.

This Job Sheet addresses the following **AST/MAST** task:

A.6. Remove and replace timing belt; verify correct camshaft timing.

Objective

Upon completion of this job sheet, you will be able to inspect the camshaft drives, including gear wear and backlash, as well as sprocket and chain wear, and replace the timing belt.

Tools and Materials

Service information	Paint stick or chalk
Belt tension gauge	Spark plug socket
Compression gauge	Remote starter switch
Flashlight	Timing belt
Large breaker bar	Scan tool
New valve or camshaft cover gasket	

Protective Clothing

Goggles or safety glasses with side shields

Describe the engine being worked on:

Year _____ Make _____ Model _____

VIN _____ Engine type and size _____

PROCEDURE

Valve Timing Check

1. Connect the scan tool to the DLC and retrieve all DTCs and related freeze-frame information. Record your finding here:

   _____

   _____

2. If any of the DTCs are valve timing, CMP, or CKP related, list what is indicated by them.

   _____

   _____

3. Conduct the specific tests given in the service information for these DTCs. Summarize what you need to do.

_____

_____

4. What were the results of the checks?

_____

_____

5. Describe the basic operation of the engine. If the timing belt or chain has slipped on the camshaft sprocket, the engine may fail to start because the valves are not properly timed in relation to the crankshaft. When the timing belt or chain has only slipped a few cogs on the camshaft sprocket, the engine has a lack of power, and fuel consumption is excessive.

_____

_____

6. To check valve timing, begin by removing the spark plug from cylinder #1. What type of ignition system does the engine have?

_____

_____

7. Disable the ignition system. How did you do this?

_____

_____

8. Connect a remote control switch to the starter solenoid terminal and the battery terminal on the solenoid. Did this connection allow you to crank the engine? If not, why not?

_____

_____

9. Place your thumb on top of the spark plug hole at cylinder #1. If this hole is not accessible, place a compression gauge in the opening. Crank the engine until compression is felt at the spark plug hole. Then, slowly crank the engine (using a breaker bar, not the remote starter) until the timing mark lines up with the zero-degree position on the timing indicator. If there is no timing indicator external to the engine, check the service information to determine the correct valve timing. Describe your findings.

_____

_____

10. Remove the rocker arm or camshaft cover and install a breaker bar and socket on the crankshaft pulley nut. Observe the valve action while rotating the crankshaft about 30 degrees before and after TDC on the exhaust stroke. In this crankshaft position, the exhaust valve should close a few degrees after TDC on the exhaust stroke, and the intake valve should open a few degrees before TDC on the exhaust stroke. What did you observe?

_____

_____

11. If the valves did not open properly in relation to the crankshaft position, the valve timing is not correct. What should you do to correct it?

_____

_____

12. If the timing was correct, reinstall the rocker arm or camshaft cover with a new gasket. Tighten the attaching bolts to the proper specification. The recommended torque is _____.

13. Reinstall the spark plug and tighten it to the proper specification. The recommended torque is _____.

Replace a Timing Belt on an OHC Engine

NOTE: *The accessory drive belt pulley on the crankshaft must be removed prior to removal of the timing belt cover.*

1. Are the camshafts on this engine driven by belts or chains? Describe the arrangement.

_____

_____

2. Disconnect the negative cable from the battery prior to beginning to remove and replace the timing belt. Install a computer memory retainer. What did you use?

_____

_____

3. Gently remove the timing cover, being careful not to distort or damage it while pulling it up. With the cover removed, check the immediate area around the belt for wires and other obstacles. If some are found, move them out of the way. What needed to be removed? Did you need to remove other drive belts?

_____

_____

_____

_____

4. Align the timing marks on the camshaft's sprocket with the mark on the cylinder head. If the marks are not obvious, use a paint stick or chalk to clearly mark them. What did you need to do?

   _____

   _____

5. Carefully remove the crankshaft timing sensor and probe holder.   Task Completed ☐

6. Loosen the adjustment bolt on the belt tensioner pulley. It is normally not   Task Completed ☐
   necessary to remove the tensioner assembly.

7. Slide the belt off the crankshaft sprocket. Do not allow the crankshaft   Task Completed ☐
   pulley to rotate while doing this.

8. To remove the belt from the engine, the crankshaft pulley may need to be removed to slip it off the crankshaft sprocket. Did you need to remove the pulley? What did you need to do in order to do this?

   _____

   _____

9. After the belt has been removed, inspect it for cracks and other damage. Cracks will become more obvious if the belt is twisted slightly. Describe any defects in the belt.

   NOTE: *Timing belts are always replaced once they have been removed.*

   _____

   _____

10. Inspect the gears for wear or cracks, and replace necessary parts. What do you need to do?

   _____

   _____

11. Is the engine equipped with variable valve timing? If so, describe the basic system. If not, proceed to step 16.

   _____

   _____

12. Describe the basic checks of the variable valve timing system as described in the service information.

   _____

   _____

13. Conduct those tests and describe the results.

   _____

   _____

14. Based on all of the preceding steps, what parts need to be replaced in addition to the timing chain(s) or belt(s)?

_____

_____

15. Place the belt around the crankshaft sprocket. Then, reinstall the crankshaft pulley. Make sure the timing marks on the crankshaft pulley are lined up with the marks on the engine block. If they are not, carefully rock the crankshaft until the marks are lined up. Describe where the timing marks are located.

_____

_____

16. With the timing belt fitted onto the crankshaft sprocket and the crankshaft pulley tightened in place, the crankshaft timing sensor and probe can be reinstalled. What is the procedure for installing the CKP?

_____

_____

17. Align the camshaft sprocket with the timing marks on the cylinder head. What is the procedure for installing the CMP? Then, wrap the timing belt around the camshaft sprocket and allow the belt tensioner to put a slight amount of pressure on the belt.

_____

_____

18. Adjust the tension as described in the service information. Then, rotate the engine two complete turns. Recheck the tension. What are the specifications for belt tension? Why do you need to rotate the engine twice before rechecking the tension?

_____

_____

19. Many manufacturers recommend that a relearn or initialization procedure be followed after anything in the valve timing system has been serviced. What does the manufacturer recommend for this engine?

_____

_____

20. Start the engine and check the DTCs and freeze-frame data. What did you find and what does this indicate?

_____

_____

Problems Encountered

_____

_____

_____

Instructor's Comments

_____

_____

_____

ENGINE REPAIR JOB SHEET 15

Repair and Replace Damaged Threads

Name _____ Station _____ Date _____

NATEF Correlation

This Job Sheet addresses the following **MLR** task:

A.6. Perform common fastener and thread repair, to include: remove broken bolt, restore internal and external threads, and repair internal threads with thread insert.

This Job Sheet addresses the following **AST/MAST** task:

A.7. Perform common fastener and thread repair, to include: remove broken bolt, restore internal and external threads, and repair internal threads with thread insert.

Objective

Upon completion of this job sheet, you will be able to inspect and repair internal and external threads, as well as install thread inserts and remove broken bolts.

Tools and Materials

HeliCoil or TIME-SERT set	Vise grips
Small cape chisel	Screw extractor set
Tap and die set	Drill
Drill bits	Center punch
Cutting fluid	Air nozzle

Protective Clothing

Goggles or safety glasses with side shields

PROCEDURE

Removing Broken Bolts

1. If part of a broken bolt is protruding from the surface, grip that part with vise grips and attempt to turn the bolt out. Were you able to do this?

   _____

   _____

   NOTE: *Applying heat with a pencil/pinhead torch to the area around the broken bolt will help loosen it.*

2. If vise grips did not work or if the bolt is broken even with or below the surface, place a small cape chisel on the broken surface of the bolt.   Task Completed ☐

3. Use a hammer to tap the chisel and bolt to turn the bolt out. Once enough of the bolt is exposed, use the vise grips to remove it. Were you able to do this?

_____

_____

4. If you still cannot get the bolt out, get a screw extractor set.   Task Completed ☐

5. With a center punch, make a dimple in the exact center of the broken bolt.   Task Completed ☐

6. With the correct size drill bit (as recommended by the manufacturer of the screw extractor set), drill through the center of the bolt to the depth recommended by the tool manufacturer. Take care not to drill too deeply and make sure the drill bit moves straight through the center of the bolt.   Task Completed ☐

7. Install the correct extractor. Most often, the extractor will need to be tapped with a hammer into the new bore; other extractors have left-hand threads and must be threaded into the bore.   Task Completed ☐

8. Turn the extractor to remove the bolt. Did the bolt come out?

_____

_____

9. After the bolt has been removed, run a thread chaser through the bolt hole to straighten and clean the threads in the bore.   Task Completed ☐

Restoring Internal Threads

1. Using a thread gauge, determine the proper thread pitch and size (diameter) of the damaged threads. Describe the threads.

_____

_____

2. Select the correct tap for the bolt hole.   Task Completed ☐

3. Apply some cutting fluid to the tap.   Task Completed ☐

4. Turn the tap slowly into the hole. Stop at about a quarter to a half turn. Then, back the tap out to keep the cutting edges of the tap clean.   Task Completed ☐

5. Continue the process until the tap hits the bottom or the required depth needed. Then, turn the tap totally out.   Task Completed ☐

6. Use compressed air to clean out the hole and use a solvent to remove the cutting oil.   Task Completed ☐

7. Run the correct size thread chaser through the threads and clean out the bore with compressed air.   Task Completed ☐

8. Explain any problems you had in doing this.

_____

_____

Restoring External Threads

1. Using a thread gauge, determine the proper thread pitch and size of the damaged threads. Describe the threads.

   _____

   _____

2. Thoroughly clean the threads that will be repaired.   Task Completed ☐

3. Apply some cutting fluid to the die.   Task Completed ☐

4. Carefully start the die over the threads. If the outer edge of the threads is damaged, it is helpful to file the sharp edges off before starting the die.   Task Completed ☐

5. Turn the die slowly over the threads. Stop at about a quarter to a half turn. Then, reverse the die to keep the cutting edges of the die clean.   Task Completed ☐

6. Continue the process until the die reaches the end of the threads. Then, turn the die off the bolt.   Task Completed ☐

7. Clean the bolt and die with compressed air and solvent.   Task Completed ☐

8. Explain any problems you had in doing this.

   _____

   _____

Installing Thread Inserts

1. Determine the correct size (diameter) and thread pitch for the bore. What is it?

   _____

   _____

2. Select the appropriate insert from the kit.   Task Completed ☐

3. Using the guidelines of the insert manufacturer, determine the drill size required to install the insert. What size will you use?

   _____

   _____

4. While making sure the drill is perfectly straight, drill the required bore to the required depth. If you are going to install a TIME-SERT, put the required chamfer on the outside of the bore.   Task Completed ☐

5. Clean out that bore with compressed air.   Task Completed ☐

6. With the correct tap, thread the inside of the bore to match the external threads of the insert.   Task Completed ☐

7. Screw the insert tightly into the bore. On HeliCoil type inserts, the tang at the top of the insert must be removed with needle-nose pliers after the insert is fully seated.   Task Completed ☐

8. Explain any problems you had in doing this.

_____

_____

Problems Encountered

_____

_____

_____

Instructor's Comments

_____

_____

_____

ENGINE REPAIR JOB SHEET 16

Servicing Power Train Mounts

Name _____ Station _____ Date _____

NATEF Correlation

This Job Sheet addresses the following **AST/MAST** task:

A.8. Inspect, remove, and replace engine mounts.

Objective

Upon completion of this job sheet, you will be able to inspect, replace, and align powertrain mounts.

Tools and Materials

Engine support fixture

Engine hoist

Basic hand tools

Protective Clothing

Goggles or safety glasses with side shields

Describe the vehicle being worked on:

Year _____ Make _____ Model _____

VIN _____ Engine type and size _____

Transmission type and model _____

PROCEDURE

1. Many shifting and vibration problems can be caused by worn, loose, or broken engine and transmission mounts. Visually inspect the mounts for looseness and cracks. Give a summary of your visual inspection.

   _____

   _____

   _____

2. Pull up and push down on the engine and transaxle case while watching the mounts. If the mounts' rubber separates from the metal plate or if the engine or transaxle moves up but not down, replace the appropriate mount. If there is movement between the metal plate and its attaching point on the frame, tighten the attaching bolts to an appropriate torque. Describe the results of doing this.

   _____

   _____

3. From the driver's seat, apply the foot brake, set the parking brake, and start the engine. Put the transmission in first gear, raise the engine speed to about 1500–2000 rpm, and gradually release the clutch pedal. Watch the torque reaction of the engine on its mounts. If the engine's reaction to the torque appears to be excessive, broken or worn drive train mounts may be the cause. Describe the results of doing this.

_____

_____

4. If it is necessary to replace the engine or transaxle mount, make sure you follow the manufacturer's recommendations for maintaining the alignment of the driveline. Describe the recommended alignment procedures.

_____

_____

_____

_____

5. When removing the engine and/or transaxle mounts, begin by disconnecting the battery's negative cable.   Task Completed ☐

6. Disconnect any electrical connectors that may be located around the mounts. Be sure to label any wires you remove to facilitate reassembly.   Task Completed ☐

7. It may be necessary to move some accessories, such as the horn, in order to service the mount without damaging some other assembly.   Task Completed ☐

8. Install the engine support fixture and attach it to an engine hoist.   Task Completed ☐

9. Lift the engine just enough to take the pressure off the mounts.   Task Completed ☐

10. Remove the bolts attaching the mounts to the frame and the mounting bracket, and then remove the mount.   Task Completed ☐

11. To install the new mount, position it in its correct location on the frame and tighten its attaching bolts to the proper torque. What is the torque specification?   Task Completed ☐

12. Install the bolts that attach the mount to its bracket. Prior to tightening these bolts, check the alignment of the mount.   Task Completed ☐

13. Once you have confirmed that the alignment is correct, tighten all loosened bolts to their specified torque.   Task Completed ☐

14. Remove the engine hoist fixture from the engine and reinstall all accessories and wires that may have been removed earlier.   Task Completed ☐

Problems Encountered

_____

_____

_____

Instructor's Comments

_____

_____

ENGINE REPAIR JOB SHEET 17

Servicing a Hybrid Vehicle's Internal Combustion Engine

Name _____ Station _____ Date _____

NATEF Correlation

This Job Sheet addresses the following **MLR** task:

A.7. Identify hybrid vehicle internal combustion engine service precautions.

This Job Sheet addresses the following **AST/MAST** task:

A.9. Identify hybrid vehicle internal combustion engine service precautions.

Objective

Upon completion of this job sheet, you will be able to describe some of the necessary precautions associated with hybrid vehicle service.

Tools and Materials

Service information

Protective Clothing

Goggles or safety glasses with side shields

Special high-voltage gloves

Special tools as required by the manufacturer.

Describe the vehicle assigned to you:

Year _____ Make _____ Model _____

VIN _____ Engine type and size _____

PROCEDURE

> **NOTE:** *Manufacturers require that technicians receive special training and equipment before being allowed to service hybrid or electric vehicles. This job sheet is not intended to substitute for factory-specific training. It is assumed that students will use service information to complete this job sheet and not be exposed to actual high-voltage hazards. If students are exposed to these hazards, the instructor must ensure that proper procedures are being followed and that the proper protective tools and clothing are available to safely complete the task.*

1. A hybrid vehicle engine may start to recharge the battery. Since you do not want the vehicle to start during routine service (such as an oil change), describe the necessary precautions to insure the engine does not start.

_____

_____

_____

2. Describe the identification of high-voltage wiring and connectors on the vehicle.

_____

_____

3. Hybrid vehicle service requires the use of special service tools and personal protective wear. Describe the personal protective wear and tools required for the vehicle you are assigned.

_____

_____

_____

4. Hybrid vehicles generally have a special disconnect for the high-voltage system. Describe the location and procedure to disconnect the high-voltage system.

_____

_____

_____

5. Describe the process of using a voltmeter to verify that the high-voltage circuit is actually disconnected.

_____

_____

_____

6. Even if the high-voltage battery is disconnected, precautions must be taken in handling the high-voltage battery and its connections. Describe the precautions necessary when handling the high-voltage battery.

_____

_____

_____

Problems Encountered

_____

_____

_____

Instructor's Comments

_____

_____

_____

ENGINE REPAIR JOB SHEET 18

Removing an Engine from a Vehicle

Name _____ Station _____ Date _____

NATEF Correlation

This Job Sheet addresses the following **AST/MAST** task:

A.10. Remove and reinstall engine in an OBDII or newer vehicle; reconnect all attaching components and restore the vehicle to running condition.

Objective

Upon completion of this job sheet, you will be able to remove the engine from a front-wheel-drive vehicle.

Tools and Materials

Engine cradle and dolly   Hoist

Drain pans   Portable crane

Battery terminal puller   Hand tools

Fender covers

Protective Clothing

Goggles or safety glasses with side shields

Describe the vehicle being worked on:

Year _____ Make _____ Model _____

VIN _____ Engine type and size _____

PROCEDURE

1. Center the vehicle on a frame contact lift/hoist.   Task Completed ☐

2. Go through the vehicle and check for any memory functions for the radio, power seats, etc. Write down all settings so they can be reset after the engine is reinstalled. What memory devices are on this vehicle?

   _____

   _____

3. Remove the battery. Which cable did you disconnect first? Why?

   _____

   _____

4. If the hood must be removed, mark the location of the hinge to the hood for reference during assembly, and then remove the hood and place it in a safe location. Where did you set it?

_____

_____

5. Drain the engine oil.   Task Completed ☐

6. Drain the engine's coolant.   Task Completed ☐

7. According to the service information, will the transaxle be removed with the engine? If so, drain its fluid.

_____

_____

8. Disconnect the air induction system and remove the air cleaner assembly.   Task Completed ☐

9. Relieve the fuel pressure at the fuel rail. How did you do this?

_____

_____

10. Disconnect the fuel supply line to the fuel rail and the fuel return line to the pressure regulator.   Task Completed ☐

11. Disconnect the throttle linkage at the throttle body (if applicable). Are there brackets for the linkage that need to be removed so the linkage can be moved out of the way?

_____

_____

12. Remove all drive belts.   Task Completed ☐

13. Attempt to remove the A/C compressor without disconnecting the refrigerant lines. Then, move it out of the way. If you need to disconnect the lines, what do you need to do before doing this? And, what should you do after the lines are disconnected?

_____

_____

14. Remove the power steering lines to the pump.   Task Completed ☐

15. Remove any other hoses or lines connected to the engine.   Task Completed ☐

16. Disconnect the heater inlet and outlet hoses.   Task Completed ☐

17. Disconnect the electric cooling fans.   Task Completed ☐

18. Remove the radiator, but make sure all electrical connections to the radiator are disconnected first. Were there electrical connections?

_____

_____

19. Label and disconnect all remaining wires and vacuum hoses.   Task Completed ☐

20. Make sure the engine ground strap is disconnected.   Task Completed ☐

21. Loosen and remove the axle shaft hub nuts.   Task Completed ☐

22. Raise the vehicle so you can comfortably work under it.   Task Completed ☐

23. Remove the wheel and tire assemblies for the front wheels.   Task Completed ☐

24. Disconnect all suspension and steering parts that need to be removed according to the service information. Index the parts so wheel alignment will be close after reassembly. What needed to be removed?

    _____

    _____

25. Remove the axle shafts.   Task Completed ☐

26. Disconnect the fuel line from the fuel tank to the engine. What did you use to plug the line and prevent fuel leakage?

    _____

    _____

27. Disconnect all lines, hoses, and electrical wiring to the transmission if the transmission is to be removed with the engine. What did you need to disconnect?

    _____

    _____

28. Disconnect all linkages to the transmission. Again, do this only if the transmission comes out with the engine. What linkages were there?

    _____

    _____

29. Remove any heat shields that may be in the way.   Task Completed ☐

30. Disconnect the exhaust pipes at the exhaust manifold.   Task Completed ☐

If the engine is to be removed from the top of the engine compartment:

1. Connect the engine sling or lifting chains to the engine. Where did you attach the chains?

    _____

    _____

2. Connect the sling to the crane and raise the crane just enough to support the engine.   Task Completed ☐

3. From under the vehicle, remove the cross member.   Task Completed ☐

4. Remove the mounting bolts for the engine at the engine and transmission mounts.   Task Completed ☐

5. From under the hood, remove all remaining mounts. Task Completed ☐

6. Raise the engine slightly to free the engine from the mounts. Task Completed ☐

7. Slowly raise the engine from the engine compartment. Guide the engine around all wires and hoses to make sure nothing gets damaged. Did you have any problems doing this?

_____

_____

8. Once the engine is cleared from the vehicle, prepare to separate it from the transaxle and mount it on an engine stand. Task Completed ☐

If the engine is to be removed from the bottom of the vehicle:

1. Position the engine cradle and dolly under the engine. Task Completed ☐

2. Adjust the pegs of the cradle so they fit into the recesses on the bottom of the engine and secure the engine. Task Completed ☐

3. Remove all engine and transmission mount bolts. Task Completed ☐

4. Slowly raise the vehicle, lifting it slightly away from the engine. Task Completed ☐

5. Check the area around the engine to make sure nothing remains that should be removed or disconnected. Task Completed ☐

6. Slowly raise the vehicle over the engine while guiding all wires and hoses out of the way. Did you have any problems doing this?

_____

_____

_____

7. Once the vehicle is clear of the engine, prepare to separate the engine from the transaxle and mount it on an engine stand. Task Completed ☐

Problems Encountered

_____

_____

_____

Instructor's Comments

_____

_____

_____

ENGINE REPAIR JOB SHEET 19

Installing an Engine into a FWD Vehicle

Name _____ Station _____ Date _____

NATEF Correlation

This Job Sheet addresses the following **AST/MAST** task:

A.10. Remove and reinstall engine in an OBDII or newer vehicle; reconnect all attaching components and restore the vehicle to running condition.

Objective

Upon completion of this job sheet, you will be able to reinstall the engine in a front-wheel-drive vehicle.

Tools and Materials

Engine cradle and dolly

Hoist

Portable crane

Hand tools

Protective Clothing

Goggles or safety glasses with side shields

Describe the vehicle being worked on:

Year _____ Make _____ Model _____

VIN _____ Engine type and size _____

PROCEDURE

1. Was the engine removed from the top or underneath? _____

If the engine will be installed through the top:

1. Connect the engine to a sling and then connect the sling to the crane. What did you connect the sling to?

   _____

   _____

2. Slowly lower the engine into the engine compartment. Guide the engine around all wires and hoses to make sure nothing gets damaged. Did you have any problems doing this?

   _____

   _____

3. As the engine approaches its position in the engine compartment, align the engine and transmission mounts. Did you have any problems doing this?

_____

_____

4. Once the mounts are aligned, lower the engine so you can install the bolts into the mounts.   Task Completed ☐

5. Raise the vehicle to a good working height.   Task Completed ☐

If the engine will be installed from under the car:

1. Install the engine onto the engine cradle and dolly. What did you use to secure the engine onto the dolly?

_____

_____

2. Lift the vehicle on a hoist or lift.   Task Completed ☐

3. Position the engine under the vehicle.   Task Completed ☐

4. Slowly lower the vehicle over the engine while guiding all wires and hoses out of the way. Did you have any problems doing this?

_____

_____

5. As the vehicle gets close to the engine, align the engine and transmission mounts. Did you have any problems doing this?

_____

_____

6. Once the mounts are aligned, lower the vehicle so you can install the bolts into the mounts.   Task Completed ☐

7. Raise the vehicle to a good working height.   Task Completed ☐

After the engine is resting on its mounts:

1. While working under the vehicle, install the axle shafts.   Task Completed ☐

2. Install the remaining engine and transaxle mounts and braces. What do these connect to?

_____

_____

_____

3. Connect the exhaust manifold to the exhaust system. Was a new gasket required? Did you have any problems doing this?

_____

_____

4. Install any heat shields that were removed when the engine was removed.   Task Completed ☐

5. Connect all linkages to the transmission. What linkages were there?

   _____

   _____

6. Connect all lines, hoses, and electrical wiring to the transmission. What did you need to connect?

   _____

   _____

7. Reconnect all suspension and steering parts that were disconnected or removed. After you do this, will the vehicle need a front-wheel alignment? Why?

   _____

   _____

8. Install the wheels and tires. What is the torque spec for the wheel lugs?

   _____

9. Tighten the axle hub nuts. What is the torque spec for these nuts?

   _____

10. Reconnect the fuel line from the fuel tank to the engine. Where is this connection made?

    _____

    _____

11. Lower the vehicle so you can work under the hood.   Task Completed ☐

12. Connect any remaining disconnected fuel lines.   Task Completed ☐

13. Connect heating system hoses. Did you need to replace any? Why?

    _____

    _____

14. Connect the engine ground strap.   Task Completed ☐

15. Connect all electrical connectors and wires.   Task Completed ☐

16. Connect all vacuum and other hoses.   Task Completed ☐

17. Connect the throttle linkage and adjust it according to the manufacturer's procedures. Did you need to adjust the linkage? If so, how did you do it?

    _____

    _____

18. Install the radiator and the cooling fan(s).   Task Completed ☐

19. Connect the rest of the hoses for the cooling system. Did you need to replace any of these? Why?

_____

_____

20. Install the air induction system.   Task Completed ☐

21. Connect any remaining items.   Task Completed ☐

22. Install the battery and connect the battery cables. Which cable did you connect first? Why?

_____

_____

23. Fill the radiator with coolant and visually check for leaks. Did you find any?

_____

_____

24. If the engine doesn't have oil in it already, add the specified amount of the proper type of oil.   Task Completed ☐

25. Prime the oil pump of the engine. How did you do that?

_____

_____

26. Prepare the engine for startup. What did you need to do?

_____

_____

Problems Encountered

_____

_____

_____

Instructor's Comments

_____

_____

_____

ENGINE REPAIR JOB SHEET 20

Remove, Inspect, and Install a Cylinder Head

Name _____ Station _____ Date _____

NATEF Correlation

This Job Sheet addresses the following **AST/MAST** tasks:

B.1. Remove cylinder head; inspect gasket condition; install cylinder head and gasket; tighten according to manufacturer's specifications and procedures.

B.2. Clean and visually inspect cylinder head for cracks; check gasket surface areas for warpage and surface finish; check passage condition.

Objective

Upon completion of this job sheet, you will be able to remove and install a cylinder head and cylinder head gasket, and tighten them according to the manufacturer's specifications. You will also be able inspect a cylinder head(s) for cracks, check the gasket surface areas for warpage and leakage, and check the condition of the coolant and water passages.

Tools and Materials

Service information   Soft-faced hammer

Feeler gauge   Straightedge

Dye penetrant or Magnaflux kit   Torque wrench inspection equipment

Feeler gauge

Torque angle gauge

Protective Clothing

Goggles or safety glasses with side shields

Describe the engine being worked on:

Year _____ Make _____ Model _____

VIN _____ Engine type and size _____

PROCEDURE

> **NOTE:** *This procedure may not work on all engines, so refer to the service information for the correct procedure for your engine. As you progress through the given procedure and find a step not applicable for your engine, write a note explaining the change.*

Removal

1. Place covers on the fenders and disconnect the negative lead at the battery.   Task Completed ☐

2. Be prepared to mark all wiring and hoses as they are removed or disconnected. Also, be sure that they do not contact other wiring or hoses, or interfere with other parts.   Task Completed ☐

3. Relieve the fuel pressure. How did you do this?

_____

_____

4. Drain and capture the engine's coolant and oil.   Task Completed ☐

5. Disconnect and plug the fuel line. Then, remove the intake air duct.   Task Completed ☐

6. Remove the drive belt.   Task Completed ☐

7. Disconnect and label the wiring from all sensors attached to the cylinder head or that may interfere with the removal of the head and manifolds. What sensors did you need to disconnect?

_____

_____

_____

_____

8. Disconnect and label all vacuum or emissions-related hoses that may be attached to the cylinder head or that may interfere with the removal of the head and manifolds. Briefly describe what you needed to disconnect.

_____

_____

_____

_____

9. Remove the intake manifold. Did the service information recommend an order to follow when loosening the mounting bolts? If so, what was it?

_____

_____

10. Remove the exhaust manifold. Did the service information recommend an order to follow when loosening the mounting bolts? If so, what was it?

_____

_____

11. Remove the upper radiator hose, heater hoses, and water bypass hose.   Task Completed ☐

12. Remove and set aside the connectors of the engine's wiring harness and its clamps from the cylinder head.   Task Completed ☐

13. Remove the timing belt (chain) cover, and then remove the belt or chain. To do this, what else did you need to remove?

_____

_____

14. Remove the rocker arm assembly (if equipped) according to the procedure outlined in the service information.   Task Completed ☐

15. Make sure the cylinder head is cool by placing your hand on it. If the head feels warm, wait before proceeding.   Task Completed ☐

16. Locate the recommended sequence for loosening the head bolts. Basically describe the recommended sequence.

    _____

    _____

17. Unscrew the bolts according to the sequence, one-third of a turn at a time. Repeat this sequence until all of the bolts are loosened. Then, remove them.   Task Completed ☐

18. Remove the cylinder head. If this is difficult, use a rubber mallet to carefully strike the side of the head. This may help break the head free. Did you have any difficulty removing the cylinder head?

    _____

19. Place the cylinder head on a workbench.   Task Completed ☐

Inspection

1. Make sure the sealing surfaces on the cylinder head are clean. Wipe them down with a clean rag.   Task Completed ☐

2. Describe the overall visual condition of the head.

    _____

    _____

3. Is the cylinder head iron or aluminum?

    _____

4. What are the specs for the maximum surface warpage for this head?

    _____

5. Position a straightedge diagonally over the head's sealing surface. Attempt to slide a feeler gauge strip of the same dimension as the maximum warpage limit between the straightedge and the cylinder head. Also, repeat the process by moving the straightedge along the edges, and each end and center of the cylinder head. Were you able to fit the feeler gauge between the straightedge and the cylinder head at any of the positions? What does this indicate?

    _____

    _____

6. If this cylinder head is distorted, what should be done to correct it?

    _____

    _____

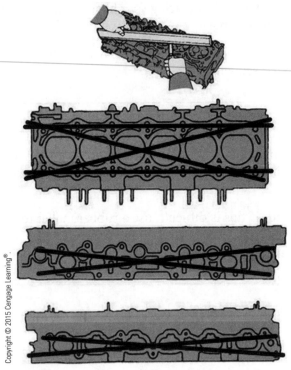

Figure 1-11 Checking the cylinder head for warpage with a straightedge and feeler gauge.

7. The cylinder head should be checked for cracks. What technique are you going to use?

_____

_____

8. Check for cracks and describe your findings.

_____

_____

9. Run a small diameter brush through all water and oil passages in the cylinder head. This will clean out the passages.

Task Completed ☐

Installation

1. Make sure the sealing surfaces on the engine block and cylinder head are clean. Wipe them down with a clean rag prior to installing the cylinder head including the threads in the block.

Task Completed ☐

2. Set the crankshaft to top dead center (TDC). Align the TDC mark on the crankshaft sprocket with the pointer on the cylinder block.

Task Completed ☐

3. Carefully look over the head bolts. Are they torque-to-yield bolts? What difference does that make during installation?

_____

_____

4. Inspect the threads of the bolts and replace any that need to be replaced. Make sure the replacement bolts are an exact duplicate of what was originally used. What are your findings?

_____

_____

5. Thoroughly clean the threads of each bolt. Does the service information recommend that the bolts be put in dry or should they have a lubricant? If yes, what lubricant?

_____

6. Sort the head bolts and identify any that are longer than the others. Are all of your head bolts the same length?

_____

7. If some bolts are longer than others, check the service information to determine their proper location. Where do the longer head bolts go in this engine?

_____

8. Install the dowel pins on the engine block.   Task Completed ☐

9. Position the cylinder head gasket onto the block and check its fit. Make   Task Completed ☐
   sure all bores and passages line up properly and that the correct side of the
   gasket is facing up.

10. Carefully lower the cylinder head over the head gasket and onto the block.   Task Completed ☐

11. Insert the cylinder head bolts into their proper bore.   Task Completed ☐

12. Start each of the bolts by hand turning them.   Task Completed ☐

13. Refer to the service information for the proper tightening sequence for the head bolts. Describe that sequence.

_____

_____

14. Check the specifications for tightening steps. If the manufacturer recommends steps, describe them here.

_____

_____

15. Tighten the bolts according to the specifications. The final amount of torque on the head bolts should be what?

_____

16. Do the specifications call for additional tightening? If so, how much?

_____

_____

17. Install the rocker arm assembly and tighten the bolts to specifications. What are the torque specifications?

_____

18. Install the cam chain or belt. Make sure the camshaft sprocket is properly aligned. Task Completed ☐

19. Reinstall the wiring harness clamps to the cylinder head. Task Completed ☐

20. Install the upper radiator hose, heater hoses, and water bypass hose. Task Completed ☐

21. Install and tighten the exhaust manifold and gasket(s). What are the torque specifications?

_____

22. Install and tighten the intake manifold and gasket(s). What are the torque specifications?

_____

23. Connect all disconnected vacuum hoses. Task Completed ☐

24. Install the fuel line. Task Completed ☐

25. Reconnect all electrical connectors to their appropriate sensors. Task Completed ☐

26. Install the intake air duct and related parts. Task Completed ☐

27. Make sure all tubes, hoses, and connectors are installed correctly. Task Completed ☐

28. Reconnect the negative battery cable. Task Completed ☐

29. Check the system for fuel leaks by turning the ignition switch ON (do not operate the starter) so that the fuel pump runs for about 2 seconds and pressurizes the fuel line. Repeat this operation two or three times, and then check for fuel leakage at any point in the fuel line. Did you have any leaks?

_____

_____

30. Refill the radiator with engine coolant, and bleed air from the cooling system with the heater valve open. Task Completed ☐

31. Add the specified amount of new oil to the engine. Task Completed ☐

32. Start the engine and check for leaks. Task Completed ☐

33. Using the service information, locate all systems that need to be reinitialized or reset, including the clock. What systems require resetting?

_____

_____

_____

Problems Encountered

_____

_____

_____

Instructor's Comments

_____

_____

_____

ENGINE REPAIR JOB SHEET 21

Inspect Pushrods and Rocker Arms

Name _____ Station _____ Date _____

NATEF Correlation

This Job Sheet addresses the following **AST/MAST** task:

B.3. Inspect pushrods, rocker arms, rocker arm pivots and shafts for wear, bending, cracks, looseness, and blocked oil passages (orifices); determine necessary action.

Objective

Upon completion of this job sheet, you will be able to inspect pushrods, rocker arms, rocker arm pivots and shafts for wear, bending, cracks, looseness, or blocked oil passages (orifices) and inspect valve lifters.

Tools and Materials

Inside and outside micrometers
Service information
Small, flat-blade screwdriver or pick
Surface plate

Protective Clothing

Goggles or safety glasses with side shields

Describe the engine being worked on:

Year _____ Make _____ Model _____

VIN _____ Engine type and size _____

PROCEDURE

1. Visually inspect each valve lifter where it contacts the camshaft lobe. If you find any signs of wear, check with your instructor. What did you find? If the lifter is a flat tappet, was the wear pattern on the bottom of the lifter normal?

   _____

   _____

2. With your finger or a pushrod, firmly press down on the plunger of the lifter. Make sure it moves freely and with some resistance. If there is no movement or if the plunger moves easily, replace the lifter. What did you find?

   _____

   _____

3. Roll each pushrod across a surface plate or drill press table. Did any of the pushrods wobble? If so, what does that indicate?

   _____

   _____

4. Inspect the oil passage of each push rod and the tips for galling and wear. Summarize your findings.

   _____

   _____

5. Visually inspect the contact area and bearing surface of each rocker arm. Summarize your findings.

   _____

   _____

6. What are your service recommendations for the rocker arms?

   _____

   _____

7. If your engine has a rocker shaft, look for evidence of wear in the area where the rocker arm fits. What did you find?

   _____

   _____

8. Use inside and outside micrometers to determine the clearance between the rocker shaft and rocker arm. Compare the measurement to specifications. What did you find?

   _____

   _____

9. If the above clearance is excessive, what should be done to correct the problem?

   _____

   _____

Problems Encountered

_____

_____

_____

Instructor's Comments

_____

_____

_____

ENGINE REPAIR JOB SHEET 22

Adjusting Valves

Name _____ Station _____ Date _____

NATEF Correlation

This Job Sheet addresses the following **MLR** task:

B.1. Adjust valves (mechanical or hydraulic lifters).

This Job Sheet addresses the following **AST/MAST** task:

B.4. Adjust valves (mechanical or hydraulic lifters).

Objective

Upon completion of this job sheet, you will be able to adjust the valves on an OHC engine.

Tools and Materials
Service information
Feeler gauge set
Remote starter button
Hand tools

Protective Clothing
Goggles or safety glasses with side shields

Describe the vehicle being worked on:
Year _____ Make _____ Model _____

VIN _____ Engine type and size _____

Describe the valve train arrangement. Are there rocker arms, hydraulic lifters, etc.?

PROCEDURE

>   **NOTE:** *This procedure does not apply to all engines and is based on a four-cylinder overhead camshaft engine with valve lash adjusting screws. Engines with different arrangements require a different procedure. Always refer to the service information for the correct procedure. Record where the manufacturer's procedure varies from the following procedure at each step.*

1. Check the service information to determine if the valves should be adjusted with a cold engine or when the engine is at normal operating temperature. Describe the recommended condition.

   _____

   _____

2. What is the specification for valve lash clearance?

   _____

   _____

3. When the engine is at the desired temperature, remove the valve (cam) cover.

   _____

   _____

4. Set the #1 piston at top dead center (TDC).   Task Completed ☐

5. Select the correct thickness feeler gauge from the set.   Task Completed ☐

6. Insert the feeler gauge between the adjusting screw and the end of the valve stem, and slide it back and forth. You should feel a slight amount of drag. What did you observe?

   _____

   _____

7. If you felt too much or too little drag, loosen the locknut with the special   Task Completed ☐
   tools, and turn the adjusting screw until the drag on the feeler gauge is correct.

8. Tighten the locknut.   Task Completed ☐

9. Recheck the valve clearance. Repeat the adjustment if necessary. Task Completed ☐

10. Check and correct the other valve(s) in cylinder #1. Did you need to adjust them?

    _____

    _____

11. Rotate the crankshaft 180° clockwise (the camshaft pulley turns 90°).   Task Completed ☐

12. Check and, if necessary, adjust the clearances of the valves on cylinder #3. Did you need to adjust them?

    _____

    _____

13. Rotate the crankshaft 180° clockwise (the camshaft pulley turns 90°).   Task Completed ☐

14. Check and, if necessary, adjust the clearances of the valves on cylinder #4. Did you need to adjust them?

    _____

    _____

15. Rotate the crankshaft 180° clockwise (camshaft pulley turns 90°).   Task Completed ☐

16. Check and, if necessary, adjust the clearances of the valves on cylinder #2. Did you need to adjust them?

_____

_____

17. Install the cylinder head cover.   Task Completed ☐

Problems Encountered

_____

_____

_____

Instructor's Comments

_____

_____

_____

ENGINE REPAIR JOB SHEET 23

Inspect and Replace Camshaft Drives and Timing Belts

Name _____ Station _____ Date _____

NATEF Correlation

This Job Sheet addresses the following **AST/MAST** tasks:

B.5. Inspect and replace camshaft and drive belt/chain (includes checking drive gear wear and backlash, end play, sprocket and chain wear, overhead cam drive sprocket(s), drive belt(s), belt tension, tensioners, camshaft reluctor ring/tone-wheel, and variable valve timing components).

B.6. Establish camshaft position sensor indexing.

Objective

Upon completion of this job sheet, you will be able to inspect the camshaft drives, including gear wear and backlash, and sprocket and chain wear.

Tools and Materials

Service information   Paint stick or chalk

Belt tension gauge   Spark plug socket

Compression gauge   Remote starter switch

Flashlight   Large breaker bar

New valve or camshaft cover gasket   Scan tool

Protective Clothing

Goggles or safety glasses with side shields

Describe the engine being worked on:

Year _____ Make _____ Model _____

VIN _____ Engine type and size _____

PROCEDURE

Valve Timing Check

1. Connect the scan tool to the DLC and retrieve all DTCs and related freeze-frame information. Record your finding here.

   _____

   _____

2. If any of the DTCs are valve timing, CMP, or CKP related, list what is indicated by them.

   _____

   _____

3. Conduct the specific tests given in the service information for these DTCs. Summarize what you need to do.

   _____

   _____

4. What were the results of the checks?

   _____

   _____

5. Describe the basic operation of the engine. If the timing belt or chain has slipped on the camshaft sprocket, the engine may fail to start because the valves are not properly timed in relation to the crankshaft. When the timing belt or chain has only slipped a few cogs on the camshaft sprocket, the engine has a lack of power, and fuel consumption is excessive.

   _____

   _____

6. To check valve timing, begin by removing the spark plug from cylinder #1. What type of ignition system does the engine have?

   _____

   _____

7. Disable the ignition system. How did you do this?

   _____

   _____

8. Connect a remote control switch to the starter solenoid terminal and the battery terminal on the solenoid. Did this connection allow you to crank the engine? If not, why not?

   _____

   _____

9. Place your thumb on top of the spark plug hole at cylinder #1. If this hole is not accessible, place a compression gauge in the opening. Crank the engine until compression is felt at the spark plug hole. Then, slowly crank the engine until the timing mark lines up with the zero-degree position on the timing indicator. What is indicated by this?

   _____

   _____

10. Remove the rocker arm or camshaft cover and install a breaker bar and socket on the crankshaft pulley nut. Observe the valve action while rotating the crankshaft about 30 degrees before and after TDC on the exhaust stroke. In this crankshaft position, the exhaust valve should close a few degrees after TDC on the exhaust stroke, and the intake valve should open a few degrees before TDC on the exhaust stroke. What did you observe?

_____

_____

11. If the valves did not open properly in relation to the crankshaft position, the valve timing is not correct. What should you do to correct it?

_____

_____

12. If the timing was correct, reinstall the rocker arm or camshaft cover with a new gasket. Tighten the attaching bolts to the proper specification. The recommended torque is _____.

13. Reinstall the spark plug and tighten it to the proper specification. The recommended torque is _____.

Replace a Timing Belt on an OHC Engine

1. Are the camshafts on this engine driven by belts or chains? Describe the arrangement.

_____

_____

2. Disconnect the negative cable from the battery prior to beginning to remove and replace the timing belt. Install a computer memory retainer. What did you use?

_____

_____

3. Carefully remove the timing cover, being careful not to distort or damage it while pulling it up. With the cover removed, check the immediate area around the belt for wires and other obstacles. If some are found, move them out of the way. What needed to be removed? Did you need to remove other drive belts?

_____

_____

_____

_____

4. Align the timing marks on the camshaft's sprocket with the mark on the cylinder head. If the marks are not obvious, use a paint stick or chalk to clearly mark them. What did you need to do?

_____

_____

5. Carefully remove the crankshaft timing sensor and probe holder.   Task Completed ☐

6. Loosen the adjustment bolt on the belt tensioner pulley. It is normally not   Task Completed ☐
necessary to remove the tensioner assembly.

7. Slide the belt off the crankshaft sprocket. Do not allow the crankshaft   Task Completed ☐
pulley to rotate while doing this.

8. To remove the belt from the engine, the crankshaft pulley may need to be removed to slip it off the crankshaft sprocket. Did you need to remove the pulley? What did you need to do in order to do this?

_____

_____

9. After the belt has been removed, inspect it for cracks and other damage. Cracks will become more obvious if the belt is twisted slightly. Describe any defects in the belt.

   NOTE: _Timing belts are always replaced once they have been removed. When may this not be true?_

_____

_____

10. Inspect the gears for wear or cracks, and replace necessary parts. What do you need to do?

_____

_____

11. Is the engine equipped with variable valve timing? If so, describe the basic system. If not, proceed to step 16.

_____

_____

12. Describe the basic checks of the variable valve timing system as described in the service information.

_____

_____

13. Conduct those tests and describe the results.

_____

_____

14. Based on all of the preceding steps, what parts need to be replaced in addition to the timing chain(s) or belt(s)?

_____

_____

15. Place the belt around the crankshaft sprocket. Then, reinstall the crankshaft pulley. Make sure the timing marks on the crankshaft pulley are lined up with the marks on the engine block. If they are not, carefully rock the crankshaft until the marks are lined up. Describe where the timing marks are located.

_____

_____

16. With the timing belt fitted onto the crankshaft sprocket and the crankshaft pulley tightened in place, the crankshaft timing sensor and probe can be reinstalled. What is the procedure for installing the CKP?

_____

_____

17. Align the camshaft sprocket with the timing marks on the cylinder head. What is the procedure for installing the CMP? Next, wrap the timing belt around the camshaft sprocket and allow the belt tensioner to put a slight amount of pressure on the belt.

_____

_____

18. Adjust the tension as described in the service information. Then, rotate the engine two complete turns. Recheck the tension. What are the specifications for belt tension? Why do you need to rotate the engine twice before rechecking the tension?

_____

_____

19. Many manufacturers recommend that a relearn or initialization procedure be followed after anything in the valve timing system has been serviced. What does the manufacturer recommend for this engine?

_____

_____

20. Start the engine and check the DTCs and freeze-frame data. What did you find and what does this indicate?

_____

_____

Problems Encountered

_____

_____

_____

Instructor's Comments

_____

_____

_____

ENGINE REPAIR JOB SHEET 24

Disassembling a Cylinder Head

Name _____ Station _____ Date _____

NATEF Correlation

This Job Sheet addresses the following **MAST** task:

B.7. Inspect valve springs for squareness and free height comparison; determine necessary action.

Objective

Upon completion of this job sheet, you will be able to disassemble a cylinder head to inspect the valves, valve seats, valve springs, and other related parts.

Tools and Materials

Service information   Small magnet

Soft-faced hammer   Valve spring compressor

Straightedge   Square

Valve spring pressure tester   Feeler gauge

Protective Clothing

Goggles or safety glasses with side shields

Describe the engine being worked on:

Year _____ Make _____ Model _____

VIN _____ Engine type and size _____

PROCEDURE

1. Place the cylinder head on a flat surface with the valve heads down. With a soft-faced hammer, tap on the valve spring retainers. Why should you do this?

   _____

   _____

2. Adjust the jaws of a valve spring compressor so they fit the spring retainers. Put the cylinder head on its side so you have access to both the valve head and the tip. What type of valve spring compressor will you be using?

   _____

   _____

3. Begin at one end of the cylinder head and remove the valve assemblies by compressing the valve spring and removing the valve locks. A small magnet makes removing the locks easier. Did you have any difficulty doing this?

_____

_____

4. Slowly release the spring compressor and remove it from the head. Collect the valve spring and spring retainer, and then pull the valve out. Sometimes the valve tip is mushroomed. This condition will make it difficult to pull out the valve. If the tip has a lip on it, it should be filed off before trying to remove the valve. Are any of the valve tips mushroomed?

_____

_____

5. Arrange the valve, spring, retainer, and locks together so they can be reassembled as a set and in the same location. How did you do this?

_____

_____

6. Clean all of the valve springs. Visually inspect the valve springs for cracks or signs of wear while cleaning them. What did you find?

_____

_____

7. Perform a squareness test: Set a spring upright against a square. Turn the spring until a gap appears between the spring and the square. Measure the gap with a feeler gauge. If the gap is greater than 0.060 inch, the spring should be replaced. What did you find?

_____

_____

8. Perform a spring pressure test: Check the service information for proper spring pressure at the test height. Use the spring tester to test each spring for proper pressure at the test height. Any spring that does not meet specifications should be replaced. What are the specifications and what did you find?

_____

_____

_____

9. Perform a freestanding height test: Line up all of the valve springs on a flat surface. Place a straightedge across the tops of the springs. Any height discrepancy will be obvious. There should be no more than 1/16-inch of variance between the heights of the valves. What did you find?

_____

_____

10. Measure the height of the shortest and the tallest valve spring and compare those measurements to the specifications. Based on the results of this measurement, you should be able to determine which other valves should be measured. Replace any spring that is not within specifications. What did you find?

_____

_____

Problems Encountered

_____

_____

_____

Instructor's Comments

_____

_____

_____

ENGINE REPAIR JOB SHEET 25

Replace Valve Stem Seals in Vehicle

Name _____ Station _____ Date _____

NATEF Correlation

This Job Sheet addresses the following **MAST** task:

B.8. Replace valve stem seals on an assembled engine; inspect valve spring retainers, locks/keepers, and valve lock/keeper grooves; determine necessary action.

Objective

Upon completion of this job sheet, you will be able to replace valve stem seals with the engine in a vehicle. You will also be able to inspect valve spring retainers, locks/keepers, and valve lock/keeper grooves; and determine if service or replacement is required.

Tools and Materials

Spark plug socket   Small magnet

Adapter to put compressed air into the cylinders   ID tags or tape

Valve spring compressor   Service information

Protective Clothing

Goggles or safety glasses with side shields

Describe the vehicle being worked on:

Year _____ Make _____ Model _____

VIN _____ Engine type and size _____

Describe the general condition:

_____

PROCEDURE

1. Remove the spark plugs from the engine and store them in the order they were in the engine. What type of ignition does the engine have?

   _____

   _____

2. Look over the engine and determine what electrical connectors and hoses must be disconnected in order to gain access to the valve or camshaft cover(s). Label the connectors so you can identify where they were during reassembly. Then, disconnect those that need to be disconnected. What did you need to disconnect?

_____

_____

3. Following the procedures given in the service information, remove the valve (camshaft) covers. Summarize what you needed to do.

_____

_____

4. Following the procedures given in the service information, remove the rocker arms or cam followers, keeping them in the order in which they are removed. Summarize what you needed to do.

_____

_____

5. Install the compressed air adapter into the first cylinder spark plug hole. What cylinder are you starting with?

_____

_____

6. Rotate the engine to top dead center on the same cylinder that you installed the adapter. How did you determine the cylinder was at TDC?

_____

_____

7. Hook a compressed air hose to the adapter. How much pressure did you apply to the cylinder?

_____

_____

8. How will the cylinder respond if the piston is not at TDC?

_____

_____

9. With compressed air holding the valves closed, use the valve spring compressor to compress the valve spring. Remove the retainer locks with the magnet. Did you have any difficulty doing this?

_____

_____

10. Remove the retainer, spring, and old valve stem seal. Carefully inspect the valve springs, retainers, and keepers for damage and wear. What did you find?

_____

_____

11. Carefully inspect the valve lock/keeper grooves in the valve stem. What did you find?

_____

_____

12. Install the new valve stem seal, being careful not to cut the seal on the end of the valve stem. What type of seal are you installing?

_____

_____

13. Reverse the removal procedure to reinstall the valve spring, retainer, and locks. What additional parts will you install? Why?

_____

_____

14. Repeat the preceding procedures for the valves on each cylinder. Remember to always put compressed air in the cylinders before removing the locks. What will be the order in which you install the other seals?

_____

_____

15. When all of the valve stem seals are replaced, reinstall the rocker arms or cam followers. Adjust the valves to the manufacturer's specifications. What are those specifications?

_____

_____

_____

_____

16. Now install the valve (camshaft) cover(s). Reinstall all of the vacuum hoses and the wires that were previously disconnected. Did you have any problems reconnecting anything?

_____

_____

Problems Encountered

_____

_____

_____

Instructor's Comments

_____

_____

_____

ENGINE REPAIR JOB SHEET 26

Checking Valve Guides

Name _____ Station _____ Date _____

NATEF Correlation

This Job Sheet addresses the following **MAST** task:

B.9. Inspect valve guides for wear; check valve stem-to-guide clearance; determine necessary action.

Objective

Upon completion of this job sheet, you will be able to inspect valve guides for wear, and check the valve guide height and stem-to-guide clearance. Based on the results of these checks, you will be able to make recommendations for further service.

Tools and Materials

Service information

Dial indicator

Micrometer

Small bore gauge

Protective Clothing

Goggles or safety glasses with side shields

Describe the engine being worked on:

Year _____ Make _____ Model _____

VIN _____ Engine type and size _____

PROCEDURE

1. Look up the specifications for valve stem-to-guide clearance. What are the specs?

   _____

   _____

2. With the valves removed from the cylinder head, place the cylinder head on a bench with the combustion chamber facing down.   Task Completed ☐

3. Install a small bore gauge into the top of one of the guides, and then expand it.   Task Completed ☐

4. Remove the bore gauge and measure it with a micrometer. What was your reading?

_____

_____

_____

5. Install the small bore gauge into the middle of one of the guides, and then expand it.

Task Completed ☐

6. Remove the bore gauge and measure it with a micrometer. What was your reading?

_____

_____

7. Install the small bore gauge into the bottom of one of the guides, and then expand it.

Task Completed ☐

8. Remove the bore gauge and measure it with a micrometer. What was your reading?

_____

_____

_____

9. Compare the three readings. What do these indicate?

_____

_____

10. Repeat the same guide measurement process on all guides and summarize the results.

_____

_____

11. Select one of the valves and measure the top, middle, and bottom of its stem. Record your results.

_____

_____

12. Repeat this process on all of the valves and summarize the results.

_____

_____

_____

13. What is the valve stem-to-guide clearance for each valve?

_____

_____

_____

_____

_____

_____

_____

14. Place a valve into its guide.   Task Completed ☐

15. Set up a dial indicator so its plunger rests on the head of a valve. Afterward, zero the indicator.   Task Completed ☐

16. Move the valve back and forth in the guide and observe the indicator.   Task Completed ☐

17. What was the total movement of the valve head?

_____

_____

18. What is indicated by this?

_____

_____

19. Repeat this process for all valves and summarize your results.

_____

_____

_____

_____

_____

_____

20. Based on all of the checks and tests, what services do you recommend be done to this head?

_____

_____

_____

Problems Encountered

_____

_____

_____

Instructor's Comments

_____

_____

_____

ENGINE REPAIR JOB SHEET 27

Inspect Valves and Valve Seats

Name _____ Station _____ Date _____

NATEF Correlation

This Job Sheet addresses the following **MAST** task:

B.10. Inspect valves and valve seats; determine necessary action.

Objective

Upon completion of this job sheet, you will be able to inspect valves and valve seats.

Tools and Materials
Service information
Machinist's rule or valve seat width scale
Prussian blue machinist's dye

Protective Clothing
Goggles or safety glasses with side shields

Describe the vehicle being worked on:
Year _____ Make _____ Model _____

VIN _____ Engine type and size _____

Describe general condition of the engine:

_____

_____

PROCEDURE

1. Clean the valve seats in the cylinder head with emery cloth or similar material to remove any residue. What did you use?

   _____

   _____

2. Carefully inspect the valve seats for signs of damage and uneven wear. Describe your findings here. (Be specific!)

   _____

   _____

   _____

3. Carefully inspect the seats in the cylinder head for cracks. Record your findings here.

_____

_____

4. Check the seats for looseness. Describe how you did this and what your findings were.

_____

_____

_____

5. Look up the specs for valve seat width in the service information. Record the specifications for the recommended width of the exhaust and intake valve seats.

_____

_____

6. Measure and record the width for each of the exhaust and intake valve seats.

_____

_____

7. Based on your inspection, what services do you recommend for the valve seats?

_____

_____

_____

8. Inspect the valves, including the valve face, tip, and valve keeper grooves. Describe what you found, and again be specific.

_____

_____

_____

9. Locate the following specifications in the service information for this engine and record them:

Intake valve face angle _____

Exhaust valve face angle _____

Intake valve minimum margin _____

Exhaust valve minimum margin _____

10. Use a 1/64" measuring scale to determine whether the margin is an acceptable width. Record your findings here:

Cylinder Number	Intake	Exhaust
_____	_____	_____
_____	_____	_____
_____	_____	_____
_____	_____	_____
_____	_____	_____
_____	_____	_____
_____	_____	_____
_____	_____	_____

11. Based on your inspection, what services do you recommend for the valves?

    _____

    _____

    _____

12. Clean and inspect the valve spring retainers and locks. Describe what you found.

    _____

    _____

    _____

13. Install the locks into the grooves on the valve stems and move the locks around in the grooves to be sure they have a proper fit. Describe what you found.

    _____

    _____

    _____

14. Based on your inspection, what services do you recommend for the valve retainers and locks?

    _____

    _____

    _____

15. Coat a valve face with Prussian blue dye or a similar compound. What did you use?

_____

_____

16. Place the valve into the appropriate seat and rotate it several times on the seat. Remove the valve from the seat and observe the marks from the dye. What do you see?

_____

_____

17. Is the contact area even? Is it in the correct location on the seat? Is the contact area the correct width?

_____

_____

_____

18. Repeat this procedure on all of the seats. What did you find?

_____

_____

19. Describe what you need to do to have the valve properly fit into the seat.

_____

_____

_____

_____

Problems Encountered

_____

_____

_____

Instructor's Comments

_____

_____

_____

ENGINE REPAIR JOB SHEET 28

Check Valve Spring Assembled Height and Valve Stem Height

Name _____ Station _____ Date _____

NATEF Correlation

This Job Sheet addresses the following **MAST** task:

B.11. Check valve spring assembled height and valve stem height; determine necessary action.

Objective

Upon completion of this job sheet, you will be able to measure the assembled height of a valve spring and the height of a valve stem.

Tools and Materials

Divider

Machinist rule

Valve spring tension tester

Service information

Protective Clothing

Goggles or safety glasses with side shields

Describe the vehicle being worked on:

Year _____ Make _____ Model _____

VIN _____ Engine type and size _____

PROCEDURE

1. Prior to installing the valve and fitting the valve springs, all other head-work should be completed.   Task Completed ☐

2. Install the valve into its proper valve guide.   Task Completed ☐

3. Install the valve retainer and keepers. Without the spring, these must be held in place by hand.   Task Completed ☐

4. While pulling up on the retainer, measure the distance between the bottom of the retainer and the spring pad on the cylinder head with a divider.   Task Completed ☐

5. Use a scale to determine the measurement expressed by the divider. The measurement was:

   _____

6. Compare this measurement with the specifications given in the service manual for installed spring height. What are the specifications for valve stem height?

   _____

   _____

7. If the measured installed height is greater than the specifications, a valve shim must be placed under the spring to correct the difference. Did you find a difference? What should you do? Show the math.

_____

_____

_____

8. Spring tension must be checked at the installed spring height; therefore, if a shim is to be used, insert it under the spring on the valve spring tension gauge.

Task Completed ☐

9. Compress the spring into the installed height by pressing down on the tester's lever.

Task Completed ☐

10. The tension gauge will reflect the pressure of the spring when compressed to the installed or valve closed height. What were your readings?

_____

_____

11. Compare your measurement to the specifications. What are the specifications? How do your measurements differ from the specifications?

_____

_____

12. Now compress the spring to the open height specification. Use the rule on the gauge or a scale to measure the compressed height. What is the compressed height and what tension did you measure at that height?

_____

_____

13. Compare this reading to the specifications. Any pressure outside the pressure range given in the specifications indicates that the spring should be replaced. What are the specifications? What should you do with the valve spring?

_____

_____

14. After the tension and height have been checked, the spring can be installed on the valve stem.

Task Completed ☐

Problems Encountered

_____

_____

_____

Instructor's Comments

_____

_____

_____

ENGINE REPAIR JOB SHEET 29

Inspect Valve Lifters

Name _____ Station _____ Date _____

NATEF Correlation

This Job Sheet addresses the following **MAST** task:

B.12. Inspect valve lifters; determine necessary action.

Objective

Upon completion of this job sheet, you will be able to inspect pushrods, rocker arms, rocker arm pivots and shafts for wear, bending, cracks, looseness, or blocked oil passages (orifices) and inspect valve lifters.

Tools and Materials

Inside and outside micrometers

Service information

Small, flat-blade screwdriver or pick

Surface plate

Protective Clothing

Goggles or safety glasses with side shields

Describe the engine being worked on:

Year _____ Make _____ Model _____

VIN _____ Engine type and size _____

PROCEDURE

1. Visually inspect each valve lifter where it contacts the camshaft lobe. If you find any signs of wear, check with your instructor. What did you find? Was the wear pattern on the bottom of the lifter normal? If the lifter is a roller type, is the roller surface free from damage?

   _____

   _____

   _____

2. Check the sides of the lifter for scoring. If scoring is present, inspect the lifter bore for damage as well. Describe your findings.

   _____

   _____

3. According to service information for your particular vehicle, should the valve lifters be primed with oil before they are installed into the engine? Describe your findings.

_____

_____

_____

Problems Encountered

_____

_____

_____

Instructor's Comments

_____

_____

_____

ENGINE REPAIR JOB SHEET 30

Checking a Camshaft and Its Bearings

Name _____ Station _____ Date _____

NATEF Correlation

This Job Sheet addresses the following **MAST** tasks:

B.13. Inspect and/or measure camshaft for runout, journal wear, and lobe wear.

B.14. Inspect camshaft bearings surface for wear, damage, out-of-round, and alignment; determine necessary action.

Objective

Upon completion of this job sheet, you will be able to inspect a camshaft for runout, journal wear, and lobe wear. You will also be able to check camshaft bearings and determine what is required to correct any problem found during the inspection of the bearings and camshaft.

Tools and Materials

Service information   Remote starter

Dial indicator   V-blocks

Micrometer

Protective Clothing

Goggles or safety glasses with side shields

Describe the vehicle being worked on:

Year _____ Make _____ Model _____

VIN _____ Engine type and size _____

PROCEDURE

NOTE: *This procedure may not apply to all engines; therefore consult your service information.*

1. To check camshaft endplay, push it toward the rear of the cylinder head or block.   Task Completed ☐

2. Zero the dial indicator against the end of the camshaft.   Task Completed ☐

3. Push the camshaft fully towards the front of the cylinder head and read the dial indicator. The reading is the endplay.   Task Completed ☐

4. If the reading is outside of specifications, what should be done?

_____

_____

5. Remove the camshaft and wipe it clean.   Task Completed ☐

6. Inspect the lobes of the camshaft. If any lobes are pitted, scored, or excessively worn, what should be done?

_____

_____

7. Measure the diameter of each camshaft journal. Record your measurements here.

_____

_____

_____

8. Clean the camshaft bearing surfaces in the cylinder head or block. Carefully inspect the bearing surfaces. What did you find?

_____

_____

9. Measure the inside diameter of each camshaft bearing surface. Compare these to the specifications. What were your results?

_____

_____

10. Take the same measurement at three different angles within the bearing surfaces. This will check for an out-of-round condition. What were your results?

_____

_____

11. The diameter of the bearings minus the diameter of the journals equals the oil clearance available for the camshaft. What are your clearances and how do they compare to specifications?

_____

_____

12. Set the camshaft on V-blocks. Position the blocks so they support the outer ends of the camshaft.   Task Completed ☐

13. Position a dial indicator so it can securely ride on the center journal of the camshaft.   Task Completed ☐

14. Zero the indicator.   Task Completed ☐

15. Slowly rotate the camshaft and observe the readings on the indicator. Describe what you observed and what it means.

_____

_____

16. Check each camshaft lobe for scoring, scuffing, a fractured surface, pitting, and signs of abnormal wear. Describe the condition of the lobes.

_____

_____

17. Place a micrometer around a camshaft lobe so it can measure from the heel to the nose of the lobe.   Task Completed ☐

18. Record the measurement for each intake and exhaust lobe.

_____

_____

_____

_____

19. If any measurement is outside specifications, what does this indicate and what should be done?

_____

_____

Measuring Camshaft Lobe Height with the Camshaft in the Engine

1. Make sure the pushrod is in the valve lifter socket.   Task Completed ☐

2. Install the dial indicator so that the cup-shaped adapter fits into the end of the pushrod and is in the same plane as the pushrod movement.   Task Completed ☐

3. Connect a remote starter switch into the starting circuit.   Task Completed ☐

4. With the ignition switch off, bump the crankshaft over until the lifter is on the base circle of the camshaft lobe. At this point, the pushrod will be in its lowest position.   Task Completed ☐

5. Set the dial indicator at zero.   Task Completed ☐

6. Continue to rotate the crankshaft slowly until the pushrod is in its fully raised position (highest indicator reading).   Task Completed ☐

7. Compare the total lift recorded on the indicator with specifications. If the lift on the lobe is below the specified service limits, the camshaft and lifters must be replaced. What did you find?

_____

_____

Problems Encountered

_____

_____

_____

Instructor's Comments

_____

_____

_____

ENGINE REPAIR JOB SHEET 31

Servicing a Crankshaft Vibration Damper

Name _____ Station _____ Date _____

NATEF Correlation

This Job Sheet addresses the following **AST/MAST** task:

C.1. Remove, inspect or replace crankshaft vibration damper (harmonic balancer).

Objective

Upon completion of this job sheet, you will be able to inspect and install the crankshaft vibration damper (harmonic balancer)

Tools and Materials

Anaerobic thread-locking compound   Plastic mallet
Harmonic balancer puller and installer
Flat-blade screwdriver
Service information
Block of wood

Protective Clothing

Goggles or safety glasses with side shields

Describe the engine being worked on:

Year _____ Make _____ Model _____

VIN _____ Engine type and size _____

PROCEDURE

1. Use service information to determine the proper procedure to remove the harmonic (crankshaft) balancer, including any special tools required. Briefly describe the procedure here.

   _____

   _____

   _____

2. Visually inspect the balancer for signs of wear on its center bore. Record your findings.

   _____

   _____

3. Inspect the rubber insulator for deterioration and twisting. Record your findings.

_____

_____

4. Based on these checks, does the balancer need to be replaced? Why or why not?

_____

_____

5. Carefully set the balancer over the snout of the crankshaft with the key and keyway aligned.   Task Completed ☐

6. Install the vibration damper (harmonic balancer) by carefully driving it on with a special installation tool. In most cases, the damper is installed until it bottoms out against the oil slinger and the timing sprocket. Check the service information to see how the damper should be installed. Describe the procedure here.

_____

_____

_____

7. Look up the torque spec for the balancer attaching bolt. Be sure to install the large washer behind the retaining bolt and tighten the balancer. The spec is:

_____

Problems Encountered

_____

_____

_____

Instructor's Comments

_____

_____

_____

ENGINE REPAIR JOB SHEET 32

Disassemble and Clean an Engine Block

Name _____ Station _____ Date _____

NATEF Correlation

This Job Sheet addresses the following **MAST** task:

C.2. Disassemble engine block; clean and prepare components for inspection and reassembly.

Objective

Upon completion of this job sheet, you will be able to disassemble an engine block, clean and inspect the components, and prepare all parts for reassembly.

Tools and Materials

Service information	Various pullers
Hand tools	Crack detector
Cleaning solution/solvent	Rod bolt protectors
Ridge reamer	

Protective Clothing

Goggles or safety glasses with side shields

Describe the vehicle being worked on:

Year _____ Make _____ Model _____

VIN _____ Engine type and size _____

PROCEDURE

1. Before engine disassembly, be sure the engine is securely bolted to an engine stand or sitting on blocks.

   Task Completed ☐

2. Look up disassembly instructions in the service manual for the specific model car and engine prior to beginning to disassemble the engine. Summarize those instructions.

   _____

   _____

3. Remove the valve cover or covers and disassemble the rocker arm components. After removing the rocker arm and pushrods, check the rocker area for sludge. Excessive buildup can indicate a poor oil change schedule and is a signal to look for similar wear patterns on other components. Record what you found.

   _____

   _____

4. Remove the pushrods and rocker arms or rocker arm assemblies and keep them in the exact order.

Task Completed ☐

5. Carefully check the lifters for a dished bottom or scratches. What did you find?

_____

_____

NOTE: *On an OHC engine, the timing belt/chain cover and camshaft drive mechanism must be removed before removing the cylinder head.*

6. Loosen the cylinder head bolts one or two turns each, working from the center of the cylinder head outward. Then, remove the bolts, again following the center-outward sequence.

Task Completed ☐

7. Lift the cylinder head off. Save the cylinder head so it can be compared to the new head gasket during reassembly.

Task Completed ☐

8. On overhead cam engines, the camshaft must be removed before the cylinder head can be disassembled. Before tearing down the cam follower assembly, draw a diagram and use a felt-tipped marker to label the parts.

Task Completed ☐

9. On overhead valve engines, remove the timing cover.

Task Completed ☐

10. Remove the harmonic balancer or vibration damper using the puller designed for the purpose.

Task Completed ☐

11. Remove the valve (camshaft) timing cover, components, and the camshaft. Support the camshaft during removal to avoid dragging lobes over bearing surfaces, which would damage bearings and lobes. Do not bump cam lobe edges, which can cause chipping.

Task Completed ☐

12. After the camshaft has been removed, visually examine it for any obvious defects—rounded lobes, edge wear, galling, and the like. Describe the condition of the camshaft here.

_____

_____

13. Remove the oil pan if it was not removed previously. Inspect the flange for distortion and straighten as necessary. What did you find?

_____

_____

14. Remove the oil pump.

Task Completed ☐

15. Clean and lay out the valve (camshaft) timing components for inspection.

Task Completed ☐

16. Inspect the sprockets for wear, cracks, and broken teeth. Inspect the timing gears for excessive backlash. Check the chain for slackness and wear. Record your findings.

_____

_____

17. Rotate each piston to bottom dead center and inspect the cylinder to see if there is a ring ridge. Use your fingernail to check for a ridge. If it catches on the cylinder wall, the ridge must be removed.   Task Completed ☐

18. Select the correct-size ridge reamer and position it in the cylinder with the piston at bottom dead center. Adjust the cutter against the cylinder walls according to the instructions furnished with the tool.   Task Completed ☐

19. Use a wrench to turn the tool in a clockwise direction.   Task Completed ☐

    WARNING: *Always wear safety goggles or glasses with side shields when completing this task.*

20. Rotate the tool around the cylinder until the ridge is removed. Be careful not to remove more than the ridge. How much of a ridge was there?

    _____

    _____

21. Repeat this operation in each of the other cylinders.   Task Completed ☐

22. Use an oily rag to remove the cuttings from each cylinder.   Task Completed ☐

23. Check all connecting rods and main bearing caps for correct position and numbering. If the numbers are not visible, use a center punch or number stamp to number them. Describe what you found.

    _____

    _____

24. Position one crankshaft throw at the bottom of its stroke.   Task Completed ☐

25. Remove the connecting rod nuts and cap. Tap the cap lightly with a soft hammer or wood block to aid in cap removal. Cover the rod bolts with protectors to avoid damage to the crankshaft journals.   Task Completed ☐

26. Carefully push the piston and rod assembly up and out with the wooden hammer handle or wooden drift and support the piston by hand as it comes out of the cylinder.   Task Completed ☐

27. With bearing inserts in the rod and cap, replace the cap (numbers on the same side) and install the nuts. Repeat the procedure for all other piston and rod assemblies.   Task Completed ☐

28. Remove the flywheel or flex plate. Mark the position of the crankshaft and flywheel or flex plate.   Task Completed ☐

29. Remove the main bearing cap bolts and main bearing caps.   Task Completed ☐

30. Carefully take out the crankshaft by lifting both ends equally to avoid bending and damage.   Task Completed ☐

31. Remove the main bearings and rear main oil seal from the block and the main bearing caps.   Task Completed ☐

32. Examine the bearing inserts for signs of abnormal engine conditions such as embedded metal particles, lack of lubrication, antifreeze contamination, oil dilution, uneven wear, and wrong or undersized bearings. Record your findings.

_____

_____

33. Carefully inspect the main journals on the crankshaft for damage.   Task Completed ☐

34. Remove all core plugs and oil gallery plugs. Describe what you needed to do to remove them.

_____

_____

35. Visually check the cylinder head and block and their parts for cracks or other damage before they are cleaned.   Task Completed ☐

36. While inspecting the parts, check to see if the engine has ever been torn down before by looking at the bearings for undersized markings on the bearings and/or crankshaft. What did you find?

_____

_____

37. After the block or cylinder head parts have been removed and disassembled, thoroughly clean them.   Task Completed ☐

38. Describe the type of sludge and the method you used to clean the parts.

_____

_____

_____

_____

Problems Encountered

_____

_____

_____

Instructor's Comments

_____

_____

_____

ENGINE REPAIR JOB SHEET 33

Engine Block Inspection

Name _____ Station _____ Date _____

NATEF Correlation

This Job Sheet addresses the following **MAST** task:

C.3. Inspect engine block for visible cracks, passage condition, core and gallery plug condition, and surface warpage; determine necessary action.

Objective

Upon completion of this job sheet, you will be able to inspect an engine block for visible cracks, check passage condition, check core and gallery plug condition, and surface warpage.

Tools and Materials

Tap set

Straightedge

Feeler gauge set

Miscellaneous wire brushes

Protective Clothing

Goggles or safety glasses with side shields

Describe the vehicle being worked on:

Year _____ Make _____ Model _____

VIN _____ Engine type and size _____

PROCEDURE

1. Place the block in a position in which you can easily clean the threaded bores. This may involve rotating the block periodically while doing so.

   Task Completed ☐

2. Run the correct-size tap through all the threaded bores. Clean the tap after each use. Did you have a hard time running the tap through any bores? If so, what was the problem?

   _____

   _____

3. Check to make sure the threads are in good condition. If they are not, they may need to be replaced. Record your findings and conclusions.

   _____

   _____

   _____

4. Carefully look all around the block for evidence of coolant and/or oil leaks. Summarize your results.

_____

_____

_____

5. Check the seal around each core and gallery plug. Are there signs of leakage? If so, what must be done?

_____

_____

_____

6. Remove the core and gallery plugs. What types of plugs are used in this engine block? What did you need to do to remove the plugs?

_____

_____

_____

7. After the plugs have been removed, inspect the sealing surface for the plugs and record their condition here.

_____

_____

8. Run a properly sized wire brush through the water and oil passages. Make sure all debris is removed.   Task Completed ☐

9. Carefully inspect the areas around all bores, threaded or non-threaded, for evidence of cracks. Summarize your results.

_____

_____

_____

_____

10. Place the straightedge diagonally across the deck surface.   Task Completed ☐

11. The amount of warpage is determined by the size of feeler gauge you can fit into the gap between the straightedge and the deck. What was the largest feeler gauge blade you could insert?

_____

12. Move the straightedge to the other diagonal on the deck surface and repeat the previous procedure. What was the largest feeler gauge blade you could insert?

_____

13. How much warpage is there on the deck surface? What do you recommend?

_____

_____

_____

14. After the block has been thoroughly inspected and cleaned, it is ready for further service and then assembly.

Task Completed ☐

Problems Encountered

_____

_____

_____

Instructor's Comments

_____

_____

_____

ENGINE REPAIR JOB SHEET 34

Measure Cylinder Bore

Name _____ Station _____ Date _____

NATEF Correlation

This Job Sheet addresses the following **MAST** task:

C.4. Inspect and measure cylinder walls/sleeves for damage, wear, and ridges; determine necessary action.

Objective

Upon completion of this job sheet, you will be able to inspect and measure cylinder walls for damage and wear.

Tools and Materials

Outside micrometer

Service information

Telescoping gauge

Protective Clothing

Safety goggles or glasses with side shields

Describe the engine being worked on:

Year _____ Make _____ Model _____

VIN _____ Engine type and size _____

PROCEDURE

1. Using the appropriate service information, look up the specifications for standard bore size, tolerances for out-of-roundness, and taper. Record these specifications on the Report Sheet for Measuring Cylinder Bore, found at the end of this job sheet.

   Task Completed ☐

2. Release the lock screw at the end of the handle of the telescoping gauge. Compress the plungers and, using the lock screw, secure them in the retracted position. Place the telescoping gauge into the bore, in an area below ring travel and at a 90-degree angle to the crankshaft.

   Task Completed ☐

3. Release the plungers by loosening the lock screw. Allow the plungers to expand until they contact the bore walls. Rock the telescoping gauge back and forth and side to side to check for the correct resistance. Lock the plungers into position using the lock screw. Carefully retract the telescoping gauge from the bore.

   Task Completed ☐

4. Use an outside micrometer to measure the distance between the two plunger faces. Record the reading obtained for each cylinder on the Report Sheet for Measuring Cylinder Bore.

Task Completed ☐

5. Measure the diameter of the cylinder at the highest point of piston ring travel. The difference between this measurement and the diameter below ring travel is the amount the cylinder is tapered. Record the amount of taper for each cylinder on the Report Sheet for Measuring Cylinder Bore.

Task Completed ☐

6. Measure for cylinder out-of-roundness by first measuring the bore at a point parallel to the piston pin, then at right angles to the piston. The difference between these measurements is the amount of out-of-roundness. Measure for out-of-roundness at three locations in each bore (top, middle, and bottom). Record the worst condition for each cylinder on the Report Sheet for Measuring Cylinder Bore.

Task Completed ☐

7. Perform a visual inspection of the block and record your findings on the Report Sheet for Measuring Cylinder Bore.

Task Completed ☐

Problems Encountered

_____

_____

_____

Instructor's Comments

_____

_____

_____

Name _____ Station _____ Date _____

REPORT SHEET FOR MEASURING CYLINDER BORE								
Cylinder No.	1	2	3	4	5	6	7	8
Standard bore size								
Actual bore size								
Taper limits								
Actual taper								
Out-of-round limits								
Actual out-of-roundness								

Visual Inspection

Cracks and scores	☐ OK	☐ Not serviceable
Core-hole plugs	☐ OK	☐ Not serviceable
Lifter bores	☐ OK	☐ Not serviceable
Water jackets	☐ OK	☐ Not serviceable
Machined surfaces	☐ OK	☐ Not serviceable
Bolt holes and threads	☐ OK	☐ Not serviceable
Oil galleries	☐ OK	☐ Not serviceable
Oil passages	☐ OK	☐ Not serviceable

Conclusions and Recommendations _____

_____

_____

ENGINE REPAIR JOB SHEET 35

Deglaze and Clean Cylinder Walls

Name _____ Station _____ Date _____

NATEF Correlation

This Job Sheet addresses the following **MAST** task:

C.5. Deglaze and clean cylinder walls.

Objective

Upon completion of this job sheet, you will be able to deglaze and clean the cylinder walls in an engine block.

Tools and Materials

Glaze breaker

Variable-speed electric drill or honing machine

Large, round stiff-bristled brush

Clean lint-free cloth

Protective Clothing

Goggles or safety glasses with side shields

Describe the vehicle being worked on:

Year _____ Make _____ Model _____

VIN _____ Engine type and size _____

PROCEDURE

1. Carefully inspect and measure the cylinder bores for surface condition, taper, and out-of-roundness. Record your findings here:

   _____

   _____

   _____

2. If the bores are within acceptable limits, the cylinder walls only need to be deglazed. What causes glaze on the cylinder walls?

   _____

   _____

3. State what you are going to use to spin the glaze breaker:

   _____

   _____

4. What grit of glaze breaker are you going to use?

_____

5. Run the glaze breaker up and down the bores.   Task Completed ☐

6. After deglazing, use plenty of hot, soapy water and a stiff-bristle brush to   Task Completed ☐
   clean the bores.

7. Wipe the walls dry with a lint-free cloth.   Task Completed ☐

8. Rinse the block with water and dry it again.   Task Completed ☐

9. Lightly coat the cylinder walls with clean, light engine oil. Why should
   you do this?

_____

_____

Problems Encountered

_____

_____

_____

Instructor's Comments

_____

_____

_____

ENGINE REPAIR JOB SHEET 36

Inspect Camshaft Bearings

Name _____ Station _____ Date _____

NATEF Correlation

This Job Sheet addresses the following **MAST** task:

C.6. Inspect and measure camshaft bearings for wear, damage, out-of-round, and alignment; determine necessary action.

Objective

Upon completion of this job sheet, you will be able to inspect and measure the camshaft bearings for wear, damage, out-of-roundness, and alignment.

Tools and Materials
Telescoping gauge
Straightedge
Micrometer
Feeler gauge
Service manual

Protective Clothing
Goggles or safety glasses with side shields

Describe the vehicle being worked on:
Year _____ Make _____ Model _____

VIN _____ Engine type and size _____

PROCEDURE

1. Carefully inspect the camshaft bearings. Are they in the block or in the cylinder head? Describe their condition.

   _____

   _____

2. Look up the spec for the ID of the camshaft bearings and record it here:

   _____

   _____

3. With a telescoping gauge and micrometer, measure the ID of each bearing and summarize your findings here:

_____

_____

4. Compare your measurements to the specs and state your conclusions.

_____

_____

5. Check the inside diameter (ID) of the bearings at 90-degree intervals. Are the ID measurements the same or are there signs of out-of-roundness? Summarize your findings.

_____

_____

6. If the bearings were found not to be out-of round, lay a straightedge across the bottom of the bearings. Check to see if the bores are aligned by attempting to put a feeler gauge under the straightedge in each of the bores. Summarize your findings here:

_____

_____

7. What service do you recommend after having done these checks?

_____

_____

Problems Encountered

_____

_____

_____

Instructor's Comments

_____

_____

_____

ENGINE REPAIR JOB SHEET 37

Measure Crankshaft Journals

Name _____ Station _____ Date _____

NATEF Correlation

This Job Sheet addresses the following **MAST** task:

C.7. Inspect crankshaft for straightness, journal damage, keyway damage, thrust flange and sealing surface condition, and visual surface cracks; check oil passage condition; measure end play and journal wear; check crankshaft position sensor reluctor ring (where applicable); determine necessary action.

Objective

Upon completion of this job sheet, you will be able to inspect the crankshaft for surface cracks and journal damage, check the oil passage condition, and measure for journal wear, in addition to inspecting and measuring the main bearings and connecting rod bearings for damage, clearance, and end play.

Tools and Materials

Outside micrometer   Light engine oil

Dial indicator   Feeler gauge set

Service information   Two pry bars

V-blocks

Protective Clothing

Safety goggles or glasses with side shields

Describe the engine being worked on:

Year _____ Make _____ Model _____

VIN _____ Engine type and size _____

PROCEDURE

1. Examine the crankshaft carefully. Check for the following:

   a. Are the vibration damper and flywheel mounting surfaces eroded or fretted?

   _____

   b. Are there indications of damage from previous engine failures?

   _____

   c. Do any of the journal diameters show signs of heat checking or discoloration from high operating temperatures?

   _____

    d. Are any of the sealing surfaces deeply worn, sharply ridged, or scored?

    _____

    e. Are there any signs of surface cracks or hardness distress?

    _____

2. What are your recommendations for the crankshaft so far?

    _____

    _____

3. The shaft should be supported by V-blocks positioned on the end main bearing journals.

    Task Completed ☐

4. Position a dial indicator at the 3 o'clock position on the center main bearing journal.

    Task Completed ☐

5. Set the indicator at zero and turn the crankshaft through one complete rotation. What did you observe?

    _____

    _____

6. Compare the bow of the crankshaft to the acceptable alignment/bow specifications. Explain what this means.

    _____

    _____

7. Using the service information, look up the specifications for standard crankshaft size and tolerances for normal wear. Record these specifications on the Report Sheet for Measuring Crankshaft Journals.

    Task Completed ☐

8. Check the crankshaft and use proper procedures to clean it before beginning the measurement process.

    Task Completed ☐

9. Using the proper size outside micrometer, check the #1 main bearing journal twice at each end of the journal, once horizontal to the crankshaft and once vertical. Record all of the measurements on the report sheet. (**Note:** If these measurements are different, the crankshaft main bearing journal is out-of-round or tapered and the crankshaft should be machined.)

    Task Completed ☐

10. Repeat step 9 for each of the remaining main bearing journals.

    Task Completed ☐

11. Measure the #1 connecting rod journal twice at each end of the journal, once horizontal to the crankshaft and once vertical. Record all of the measurements on the report sheet. (**Note:** If these measurements are different, the crankshaft connecting rod journal is out-of-round or tapered, and the crankshaft should be machined.)

    Task Completed ☐

12. Repeat step 11 for each of the remaining connecting rod journals.

    Task Completed ☐

13. Compare the measurements of the crankshaft journals with the specifications. If the measurements are within specifications, the crankshaft can be reinstalled in the engine. If the measurements are not within factory specifications, the crankshaft must be machined before it is reinstalled in the engine.

Task Completed ☐

Check Crankshaft End Play

1. Following the manufacturer's recommended procedures presented in the service information, install the rear main oil seal in the block and bearing cap.

Task Completed ☐

2. Thoroughly lubricate all bearings and the rear main oil seal with engine oil. Install the main bearing halves into the main bearing bores in the block. Make sure the oil holes in the bearing halves line up with the oil holes in the block.

Task Completed ☐

3. Carefully install the crankshaft into the block.

Task Completed ☐

4. Install the main bearing caps with bearings, being careful to match the location numbers on the cap with the block.

5. Tighten the main bearing cap bolts one at a time in three steps until the full, specified torque is obtained. Do not torque the thrust bearing cap at this time.

Task Completed ☐

6. Pry the crankshaft back and forth to align the thrust surfaces at the thrust bearing. While prying the crankshaft forward, pry the thrust bearing cap rearward. When assured that the surfaces are aligned, torque the cap bolts to specifications in three steps.

Task Completed ☐

7. Using the service information, locate the minimum and maximum crankshaft end-play specifications. Record specifications.

   Minimum specification _____

   Maximum specification _____

8. Pry the crankshaft toward the front of the engine. Install a dial indicator and bracket so that its plunger rests against the crankshaft flange and the indicator axis is parallel to the crankshaft axis.

Task Completed ☐

   NOTE: *This procedure can be accomplished using a feeler gauge if a dial indicator is not available. This check would be made between the thrust bearing and the thrust surfaces on the crankshaft.*

9. Set the dial indicator at zero. Pry the crankshaft rearward. Note the reading on the dial indicator. _____

10. Record your dial indicator reading: _____

11. If your results are not within the manufacturer's specifications, list your conclusions and recommendations. _____

_____

Problems Encountered

_____

_____

_____

Instructor's Comments

_____

_____

_____

Name _____ Station _____ Date _____

REPORT SHEET FOR MEASURING CRANKSHAFT MEASUREMENTS							
Main Journal Number	1	2	3	4	5	6	7
Standard journal size							
Actual journal size							
Taper limits							
Actual taper							
Out-of-round limits							
Actual out-of-roundness							
Connecting Rod Journal Number	1	2	3	4	5	6	
Standard journal size							
Actual journal size							
Taper limits							
Actual taper							
Out-of-round limits							
Actual out-of-roundness							
Conclusions and Recommendations _____							
_____							
_____							

ENGINE REPAIR JOB SHEET 38

Inspecting Engine Bearings

Name _____ Station _____ Date _____

NATEF Correlation

This Job Sheet addresses the following **MAST** tasks:

C.8. Inspect and measure main and connecting rod bearings for damage and wear; determine necessary action.

C.9. Identify piston and bearing wear patterns that indicate connecting rod alignment and main bearing bore problems; determine necessary action.

Objective

Upon completion of this job sheet, you will be able to inspect and measure main bearings for damage, clearance, and end play. In addition, you will be able to identify piston and bearing wear patterns that indicate connecting rod misalignment and main bearing bore problems.

Tools and Materials

Service information	Torque wrench
Clean lint-free rag	Plastigage
Feeler gauge	Dial indicator
Telescoping gauge set	Micrometer

Protective Clothing

Goggles or safety glasses with side shields

Describe the engine being worked on:

Year _____ Make _____ Model _____

VIN _____ Engine type and size _____

PROCEDURE

General Bearing Inspection

1. Describe the probable causes for the following bearing failures. To answer these, use your text and common sense.

   a. The surface of the bearing has cracks and/or there is a loss of lining material.

   _____

   _____

b. Fretting, highly polished areas, is evident on the back of the bearing.

_____

_____

c. The surface of the bearing has areas where material has been removed/eroded.

_____

_____

d. The bearing surface appears wiped with darkening around the edges in the central region of the bearing.

_____

_____

e. The bearing shows evidence of foreign particles embedded into the surface.

_____

_____

f. The bearing surface appears smeared with melted and redistributed bearing material.

_____

_____

g. The bearing surface is scored.

_____

_____

Inspecting Main Bearings

1. Carefully examine both sides of all the main bearings. Clean each bearing. Describe the wear pattern of each main bearing.

_____

_____

2. Describe the probable causes for the following main bearing wear patterns.

a. Excessive wear on the edges of the bearing shell.

_____

_____

b. Wear in different locations on all bearings.

_____

_____

   c. Wear patterns on the parting edges of the bearing.

_____

_____

   d. Fine scoring into the bearing.

_____

_____

   e. Metal particles embedded into the bearing surface.

_____

_____

   f. Excessive wear on the lower bearing.

_____

_____

3. Based on your examination of the main bearings, what service(s) do you recommend?

_____

_____

Inspecting Connecting Rod Bearings

1. Wipe the bores of the connecting rods down with a clean cloth, and then carefully inspect the big end bearings in each connecting rod and summarize your findings.

_____

_____

2. Carefully inspect the small end bearings in each connecting rod and summarize your findings.

_____

_____

3. Did either end of any bearing show signs of wear on one side only? What would this indicate?

_____

_____

4. With the telescoping gauge and a micrometer, measure the bore of the big end of the rod with the bearing in place. Begin by measuring the ID at both sides and in the center of the rod. Do this to each rod and summarize your findings.

_____

_____

5. With the telescoping gauge and a micrometer, measure the bore of the small end of the rod with the bearing in place. Begin by measuring the ID at both sides and in the center of the rod. Do this to each rod and summarize your findings.

_____

_____

6. Check the ID of the big end of each rod with the bearing in place. Place the telescoping gauge in the center of the bore and then measure at 90-degree intervals. Summarize your findings.

_____

_____

7. Check the ID of the small end of each rod with the bearing in place. Place the telescoping gauge in the center of the bore and then measure at 90-degree intervals. Summarize your findings.

_____

_____

8. What did the preceding checks look for?

_____

_____

9. Look up the specs for the correct ID of both connecting rod bores for this engine. Record them here.

_____

_____

10. Carefully inspect the bores of each connecting rod without the bearings and summarize your findings.

_____

_____

11. Based on all of these checks, what are your recommendations for service to the connecting rods?

_____

_____

_____

_____

Main Bearing Clearance

1. Mount the engine block upside down on an engine stand.   Task Completed ☐

2. Install main bearings into bores, being careful to properly seat them.   Task Completed ☐

3. Wipe the bearings with a clean lint-free rag.   Task Completed ☐

4. Carefully install the crankshaft into the bearings. Try not to allow the   Task Completed ☐
   crankshaft to move on the bearing surfaces.

5. Wipe the crankshaft journals with a clean rag.   Task Completed ☐

6. Place a piece of plastigage on the journal. The piece should fit between the
   radii of the journal. What color plastigage did you use?

   _____

   _____

7. Install the main caps and bearings in their proper locations and directions. Task Completed ☐
   Wipe the threads of the cap bolts with a clean rag.

8. Install the cap bolts and tighten them according to the manufacturer's
   recommendations. What are those recommendations?

   _____

   _____

9. Remove the main caps and observe the spread of the plastigage. If the   Task Completed ☐
   gage did not spread, try again with a larger gage.

10. Compare the spread of the gage with the scale given on the plastigage con-
    tainer. Compare the clearance with the specifications. Give your summary
    here.

    _____

    _____

11. Carefully scrape the plastigage off the journal surface.   Task Completed ☐

12. Wipe the journal clean with a rag.   Task Completed ☐

13. If the clearance was within the specifications, remove the crankshaft and   Task Completed ☐
    apply a good coat of fresh engine oil to the bearings.

14. Reinstall the crankshaft and apply a coat of oil to the journal surfaces.   Task Completed ☐

15. Reinstall the main caps and tighten according to specifications. What are
    those specifications?

    _____

    _____

Inspecting Pistons

1. Wipe each piston and carefully inspect it while cleaning it. Describe the
   general appearance of each piston.

   _____

   _____

2. Closely examine all piston skirts for unusual wear patterns that may indicate a twisted rod. What did you find?

_____

_____

3. What problem would be indicated by spots and/or small dents on the head of a piston, or streaks of wear on the skirt of the piston?

_____

_____

4. What problem would be indicated by a highly burnt surface along with apparent melted metal around the edges of the piston head?

_____

_____

5. What problem would be indicated by a broken or damaged piston skirt?

_____

_____

6. What problem would be indicated by streaks of wear down the sides of the piston skirt?

_____

_____

Problems Encountered

_____

_____

_____

Instructor's Comments

_____

_____

_____

ENGINE REPAIR JOB SHEET 39

Inspecting and Measuring Pistons

Name _____ Station _____ Date _____

NATEF Correlation

This Job Sheet addresses the following **MAST** tasks:

    C.10. Inspect and measure piston skirts and ring lands; determine necessary action.

    C.11. Determine piston-to-bore clearance.

    C.12. Inspect, measure, and install piston rings.

Objective

Upon completion of this job sheet, you will be able to measure and inspect pistons to determine whether they can be used again and to identify other engine problems. You will also be able to properly fit the piston in its bore and properly install the piston rings.

Tools and Materials

Service information	Scraper
Cold tank	Ring groove cleaner
Various sized micrometers	Feeler gauge set
Ring expander	

Protective Clothing

Goggles or safety glasses with side shields

Describe the vehicle being worked on:

Year _____ Make _____ Model _____

VIN _____ Engine type and size _____

PROCEDURE

Inspection

1. If the piston has not been cleaned, do so by scraping off the carbon from the top of the piston and soaking the piston in a cold tank. Describe the basic condition of the pistons prior to cleaning.

    _____

    _____

2. Clean the piston ring grooves with a ring groove cleaner. Did you have difficulty doing this? Are the grooves in poor shape?

    _____

    _____

3. Visually inspect the head of the piston and describe its condition.

   _____

   _____

4. Inspect the ring land and groove area for nicks and cracks. Describe your findings.

   _____

   _____

5. Inspect the skirt for scuffing and other damage. Describe your findings.

   _____

   _____

6. Inspect the pin boss areas and record their condition.

   _____

   _____

7. Refer to the service information and record the manufacturer's recommended procedure for measuring the piston. Describe that procedure.

   _____

   _____

8. Measure the diameter of the piston and record your measurement.

   _____

   _____

9. Measure the bore of the cylinder that will be fitted with the piston measured in step 8. Record that measurement.

   _____

   _____

10. Subtract the size of the piston from the measured bore of the cylinder. What is the clearance?

    _____

    _____

11. Compare your measured clearance to the specifications. What does this indicate?

    _____

    _____

12. Install a new piston ring backward in its appropriate groove. With a feeler gauge, measure the clearance between the ring and the top of the groove. What was that measurement?

    _____

    _____

13. Compare your measurement to the specifications. What is the specification?

_____

_____

14. What is indicated by the measured clearance?

_____

_____

15. Measure the depth of the groove by placing a new ring in its appropriate groove. Roll the ring around the groove and observe the depth of the ring in the groove. What did you observe?

_____

_____

_____

16. Refer to the service information and find the specification for the ring end gap. What is it?

_____

_____

17. Place a new ring into the cylinder bore it will be installed in. Then, square the ring in the bore by placing the appropriate piston upside down into the cylinder. Make sure the ring is at the position specified by the manufacturer. Describe the specified location for measuring the ring end gap.

_____

_____

18. With a feeler gauge, measure the gap between the ends of the piston ring. What was your measurement?

_____

_____

19. Compare that measurement to specifications. Is service required? If so, what?

_____

_____

20. After the rings have the correct end gap, gather the oil control rings. Identify the proper spacing for the piston rings. Typically, the sequence for installing the oil control rings requires these steps.

   a. Install the expander ring into the lower ring groove in the piston.

   b. Rotate the ring so that its ends are 90° from the piston pin.

   c. Install the upper oil ring side rail with its gap 45° from the end of the expander.

   d. Install the lower side rail and locate its end gap 180° from the end gap of the top rail.

Does the manufacturer recommend a different sequence than this? If so, what is it?

_____

_____

21. Carefully inspect the new compression rings for markings that indicate direction. Were there any? If so, what do the markings indicate?

_____

_____

22. Install the compression rings using this sequence:

   a. With the correct side of the ring facing up, insert the bottom compression ring into the ring expander.

   b. Expand the ring just enough to fit it over the piston. Be careful not to overstretch the ring or damage the piston while installing the ring.

   c. Install the ring into its groove.

   d. With the correct side of the ring facing up, insert the top compression ring into the ring expander.

   e. Expand the ring just enough to fit it over the piston. Be careful not to overstretch the ring or damage the piston while installing the ring.

   f. Install the ring into its groove.

   Did you have difficulties doing this?

   _____

   _____

23. Stagger the ring end gaps according to specifications. Describe how the end gaps should be arranged.

   _____

   _____

Problems Encountered

_____

_____

_____

Instructor's Comments

_____

_____

_____

ENGINE REPAIR JOB SHEET 40

Checking Balance Shafts

Name _____ Station _____ Date _____

NATEF Correlation

This Job Sheet addresses the following **MAST** task:

C.13. Inspect auxiliary shaft(s) (balance, intermediate, idler, counterbalance or silencer); inspect shaft(s) and support bearings for damage and wear; determine necessary action; reinstall, and time.

Objective

Upon completion of this job sheet, you will be able to inspect the auxiliary shaft(s) (balance, intermediate, idler, counterbalance, or silencer), as well as inspect the support bearings for damage and wear.

Tools and Materials

Micrometer

Telescoping gauge

Service information

Protective Clothing

Goggles or safety glasses with side shields

Describe the vehicle being worked on:

Year _____ Make _____ Model _____

VIN _____ Engine type and size _____

PROCEDURE

1. What is the purpose of the auxiliary shaft found in the engine you are working on?

   _____

   _____

2. With the shaft removed from the engine, carefully inspect its condition. Pay close attention to bearing journals. Summarize your findings.

   _____

   _____

3. Measure the outside diameter (OD) of the bearing journals and compare them to the specifications given in the service information. Summarize your findings.

   _____

   _____

4. Visually inspect the bearings for the shaft and run your finger over the bearing surface to check for scoring or other damage. Summarize your findings.

_____

_____

5. With a telescoping gauge, measure the ID of the bearings and record your findings.

_____

_____

6. Check the bearings for out-of-roundness and taper by measuring the bearing ID at 90-degree intervals and by measuring the bearings at their center and outside edges. Do your measurements indicate that the bearing is round and not tapered?

_____

_____

7. Subtract the OD or the bearing journals from the ID of the bearings. This is the oil clearance. What is your result?

_____

_____

8. Compare your measured oil clearance to the specs and summarize your results.

_____

_____

9. Based on these checks, what services do you recommend?

_____

_____

Problems Encountered

_____

_____

_____

Instructor's Comments

_____

_____

_____

ENGINE REPAIR JOB SHEET 41

Reassemble an Engine

Name _____ Station _____ Date _____

NATEF Correlation

This Job Sheet addresses the following **AST/MAST** task:

C.1. Remove, inspect, or replace crankshaft vibration damper (harmonic balancer).

This Job Sheet addresses the following **MAST** task:

C.14. Assemble engine block.

Objective

Upon completion of this job sheet, you will be able to inspect and install the crankshaft vibration damper (harmonic balancer) and reassemble the engine and mount its components using the correct gaskets and sealants.

Tools and Materials

Light engine oil

Anaerobic thread-locking compound

Ring compressor tool

Feeler gauges

Flat-blade screwdriver

Straightedge

Block of wood

Various drivers

Plastic mallet

Plastigage

Rubber or aluminum protectors or guides

Service information

Hydraulic press

Required gaskets and sealants

Protective Clothing

Goggles or safety glasses with side shields

Describe the engine being worked on:

Year _____ Make _____ Model _____

VIN _____ Engine type and size _____

PROCEDURE

WARNING: *Make sure every sealant you use on today's engines is oxygen-sensor safe.*

1. Visually inspect the bolts. Threads must be clean and undamaged. Discard all bolts that are not acceptable. Describe the condition of the bolts.

_____

_____

2. Gather new gaskets for the engine. Never reuse old gaskets. Even if the old gasket appears to be in good condition, it will never seal as well as a new one. Protect the new gaskets by keeping them in their packages until it is time to install them.

Task Completed ☐

3. Make sure all surfaces are free of dirt, oil deposits, rust, old sealer, and gasket material.

Task Completed ☐

4. Place rubber or aluminum protectors or guides over the threaded section of the rod bolts. Lightly coat the piston, rings, cylinder wall, crankpin, and compressor tool with light engine oil. Do not coat the rod bearings with oil at this time.

Task Completed ☐

5. Be sure that the ring gaps are located on the piston in the positions recommended in the service information.

Task Completed ☐

6. Expand the compressor tool around the piston rings. Position the steps on the compressor tool downward. Tighten the compressor tool with an Allen wrench to compress the piston rings.

Task Completed ☐

7. Rotate the crankshaft until the crankpin is at its lowest level (BDC). Place the piston/rod assembly into the cylinder bore until the steps on the compressor tool contact the cylinder block deck. Make sure the piston reference mark is in the correct relation to the front of the engine.

Task Completed ☐

8. Lightly tap on the head of the piston with a mallet handle or block of wood until the piston enters the cylinder bore.

Task Completed ☐

9. Push the piston down the bore while making sure the connecting rod fits into place on the crankpin. Remove the protective covering from the rod bolts.

Task Completed ☐

10. Look in the service information for the specs on the oil clearance for connecting rod bearings and the torque specs for the rod cap bolts. What are these specs?

_____

_____

11. Check the rod bearing oil clearances on all rods with Plastigage. Record your results.

_____

_____

12. Based on the preceding steps, what do you recommend?

_____

_____

13. If the clearances are within specs, remove all cap bolts and push the piston and rod assembly into the appropriate cylinder bore. Lubricate the bearings and crankpins.

Task Completed ☐

14. Seat the connecting rod yoke onto the crankshaft.

Task Completed ☐

15. Position the matching connecting rod cap onto each rod and finger-tighten the rod nuts. When doing this, what do you need to check?

_____

_____

16. Gently tap each cap with the plastic mallet to seat it against the crankshaft and connecting rod. Task Completed ☐

17. Torque the rod cap bolts to specs. What are those specs?

_____

18. Look in the service information for the connecting rod side clearance specs. Record them here.

_____

_____

19. Measure the side clearance of each rod with a feeler gauge and record your findings.

_____

_____

20. Compare your measurements to specs and state your service recommendations.

_____

_____

21. Check the condition of the balancer shaft or crankshaft pulley hub. Make sure the surface is smooth. If the surface is not smooth, the seal will not be able to seal. What can be done if the pulley hub has a groove in it?

_____

_____

22. Before installing the oil pan and gasket, check the flanges for warpage. Use a straightedge or lay the pan, flange side down, on a flat surface with a flashlight underneath it to spot uneven edges. Carefully check the flange around bolt holes. What were the results of this check? If the flange is distorted, what should you do?

_____

_____

23. Once it has been determined that the flanges are flat, install the oil pan with a new gasket. Task Completed ☐

24. When replacing the timing cover, remove the old gaskets and seals from the timing cover and engine block. Task Completed ☐

25. Install a new crankshaft seal using a press, seal driver, or hammer, and a clean block of wood. When installing the seal, be sure to support the cover underneath to prevent damage. Task Completed ☐

26. If the timing cover extends over the front lip of the oil pan, the front portion of the oil pan gasket will be exposed. With a sharp knife or razor blade, carefully cut off the front exposed portion of the oil pan gasket. Did you need to do this?

_____

_____

27. Apply a light coating of adhesive or sealant on the timing cover and position the gasket on the cover. Finally, mount the timing cover and torque the bolts to specifications. Check the service information to see what type of adhesive or sealant you should use. Record the manufacturer's recommendations.

_____

_____

28. If all clearances are correct, prepare to install the harmonic balancer to the crankshaft.   Task Completed ☐

29. Visually inspect the balancer for signs of wear on its center bore. Record your findings.

_____

_____

30. Inspect the rubber insulator for deterioration and twisting. Record your findings.

_____

_____

31. Based on these checks, does the balancer need to be replaced? Why or why not?

_____

_____

32. Carefully set the balancer over the snout of the crankshaft with the key and keyway aligned.   Task Completed ☐

33. Install the vibration damper (harmonic balancer) by carefully pounding on it, or using a special installation tool. In most cases, the damper is installed until it bottoms out against the oil slinger and the timing sprocket. It is best to stand the engine block on end and support the crankshaft if the damper must be pounded on. Check the service information to see how the damper should be installed. Describe the procedure here.

_____

_____

_____

34. Look up the torque spec for the balancer attaching bolt. Be sure to install the large washer behind the retaining bolt and tighten the balancer. The spec is:

_____

35. Before installing the valve or cam cover, make sure the cover's sealing flange is flat, and then apply contact adhesive to the valve cover's sealing surfaces in small dabs. Mount the valve cover gasket on the valve cover and align it in position. If the gasket has mounting tabs, use them in tandem with the contact adhesive. Allow the adhesive to dry completely before mounting the valve cover on the cylinder head. Torque the mounting bolts to specifications. The specifications are:

_____

36. The intake manifold gasket seals the joint between the intake manifold and the cylinder head. To be sure the gasket will seal, check its fit before installing it. On steel shim-type gaskets, it is necessary to put a thin and even coat of positioning sealant around the vacuum port openings and a small bead of RTV silicone around the coolant openings. Install the intake manifold with a new gasket and tighten the fastening bolts to the recommended torque specification. The specifications are:

_____

37. Install the thermostat and water outlet housing. Install the thermostat with the temperature sensor facing into the block. Make sure the gasket is positioned properly. Use the sealant recommended by the manufacturer. Take care not to tighten the housing unevenly. Tighten each mounting bolt a little at a time and tighten to specifications. The specifications are:

_____

38. Exhaust manifolds may or may not require a gasket; check the service manual. Also check the service information for the tightening sequence and torque specifications. Record this information here.

_____

_____

39. Place the exhaust manifold into position. Tighten the bolts in the center of the manifold first to prevent cracking it. If there are dowel holes in the exhaust manifold that align with dowels in the cylinder head, make sure that these holes are larger than the dowels.　　Task Completed ☐

40. Reinstall the engine sling to remove the engine from the engine stand.　　Task Completed ☐

41. Raise the engine into the air on a suitable hoist, and remove the engine stand mounting head.　　Task Completed ☐

42. Set the assembled engine on the floor and support it with blocks of wood while attaching the flywheel or flex plate.　　Task Completed ☐

43. Make sure you use the right flywheel bolts and lock washers. These bolts have very thin heads and the lock washers are thin. Make sure that the bolts are properly torqued and tightened in the correct sequence. The torque specifications are:

_____

_____

44. If the vehicle has a manual transmission, install the clutch. Make sure the transmission's pilot bushing or bearing is in place in the rear of the crankshaft and that it is in good condition. Does the vehicle have a manual transmission? What is the condition of the plot bearing or bushing?

_____

_____

_____

_____

45. Assemble the clutch onto the flywheel. Start the bolts by hand. Make sure the disc is installed in the right direction. There should be a marking on it that says "flywheel side."

Task Completed ☐

46. Using a clutch-aligning tool, align the clutch disc. Then, tighten the disc and pressure plate to the flywheel. Tighten the bolts in the proper sequence and to the correct torque. The specifications are:

_____

_____

47. On cars equipped with automatic transmissions, install the torque converter, making sure it is correctly engaged with the transmission's front pump. The drive lugs on the converter should be felt engaging the transmission front pump gear.

Task Completed ☐

48. The motor mount bolts may now be installed loosely on the block. The bolts are left loose during engine installation so that the mounts can be easily aligned with the front mount brackets. Check the condition of the mounts and describe your findings here.

_____

_____

_____

_____

Problems Encountered

_____

_____

_____

Instructor's Comments

_____

_____

_____

ENGINE REPAIR JOB SHEET 42

Testing the Coolant System

Name _____ Station _____ Date _____

NATEF Correlation

This Job Sheet addresses the following **MLR** tasks:

C.1. Perform cooling system pressure and dye tests to identify leaks; check coolant condition and level; inspect and test radiator, pressure cap, coolant recovery tank, and heater core; determine necessary action.

C.4. Inspect and test coolant; drain and recover coolant; flush and refill cooling system with recommended coolant; bleed air as required.

This Job Sheet addresses the following **AST/MAST** tasks:

D.1. Perform cooling system pressure and dye tests to identify leaks; check coolant condition and level; inspect and test radiator, pressure cap, coolant recovery tank, and heater core; determine necessary action.

D.4. Inspect and test coolant; drain and recover coolant; flush and refill cooling system with recommended coolant; bleed air as required.

Objective

Upon completion of this job sheet, you will be able to perform cooling system, cap, and recovery system pressure, combustion leakage, and temperature tests, as well as test, drain, and recover coolant, and flush and refill the cooling system with the recommended coolant, and also bleed the air as required.

Tools and Materials

Thermometer	Dye penetrant
Clean cloth rags	Cooling system tester
Black light	Coolant hydrometer
Combustion leak detector	Drain pan
Cooling system flusher	Coolant recycler

Protective Clothing

Goggles or safety glasses with side shields

Describe the vehicle being worked on:

Year _____ Make _____ Model _____

VIN _____ Engine type and size _____

Describe general condition:

_____

PROCEDURE

1. Remove the radiator cap from a cool radiator.　　Task Completed ☐

2. Insert a thermometer into the coolant.　　Task Completed ☐

3. Start the engine and let it warm up.　　Task Completed ☐

4. Watch the thermometer and the surface of the coolant.　　Task Completed ☐

5. When the coolant begins to flow, it indicates the thermostat has started to open. At what temperature did this occur?

   _____

   _____

6. What are the specifications for thermostat opening temperature?

   _____

   _____

7. Based on this check, what is the condition of the thermostat?

   _____

   _____

8. Turn off the engine and allow it to cool.　　Task Completed ☐

9. Wipe off the radiator filler neck and inspect it.　　Task Completed ☐

   Is the sealing seat free of accumulated dirt, nicks, or anything that might prevent a good seal?　　☐ Yes　　☐ No

10. Inspect the tube from the radiator to the expansion tank for dents and other obstructions. Run wire through the tube to be certain it is clear.　　Task Completed ☐

11. The source of external coolant leaks can be found by a thorough visual inspection or through the use of a cooling system dye. Visually, the point of the leak may be wet or have a light grey color. The latter is the result of the coolant evaporating at that point. Are there any signs of leakage?

    _____

    _____

12. Gather the coolant dye and black light.　　Task Completed ☐

13. Pour the dye into the cooling system.　　Task Completed ☐

14. Run the engine until it reaches operating temperature.　　Task Completed ☐

15. Turn the engine off.　　Task Completed ☐

16. Inspect the engine and cooling system with the black light. A bright or fluorescent green color will be seen wherever the dyed coolant has leaked. Were there any leaks?

    _____

    _____

17. To test the cooling system for external leaks, attach a cooling system tester   Task Completed ☐
    to the appropriate flexible adapter and carefully pump up the pressure.
    Looking at the tester's gauge, bring the pressure up to the proper cooling
    system test point indicated on the dial face.

    Does the pressure begin to drop?   ☐ Yes ☐ No

    If so, check all external connections, including hoses, gaskets, and the   Task Completed ☐
    heater core for leaks.

18. Check all points closely for small pinhole leaks or potentially dangerous   Task Completed ☐
    weak points in the system.

To test for internal leaks:

19. **Ensure the engine is cool.** Remove the coolant pressure cap.   Task Completed ☐

20. Start the engine and allow it to reach normal operating temperature so the   Task Completed ☐
    thermostat will open fully.

    WARNING: *The pressure buildup from a combustion leak into the cooling system can rapidly
    build pressure in the cooling system, even if the engine is cool. It is always a good
    idea to squeeze the upper radiator hose before removing a pressure cap. If the upper
    radiator hose feels very solid, chances are the system is under pressure, even if the
    engine is cold.*

21. Slowly and carefully remove the pressure cap and replace it with the flex-   Task Completed ☐
    ible adapter.

    WARNING: *The engine coolant will be under pressure. Always wear
    hand protection and safety goggles or glasses with side
    shields when performing this task.*

22. Lock the tester slowly into place. Start the engine. Watch for pressure   Task Completed ☐
    buildup.

    WARNING: *If the pressure builds up suddenly due to an internal leak, turn off the engine, allow
    the engine to cool down and the pressure to bleed off before working on the vehicle.
    Perform a cylinder leakage test to locate the source of the combustion leak into the
    cooling system.*

23. If no immediate pressure buildup is visible on the gauge, keep the engine   Task Completed ☐
    running. Pressurize the cooling system.

24. Does the gauge fluctuate? If so, the cooling system has a compression or   ☐ Yes ☐ No
    combustion leak.

25. Before removing it, hold the tester body and press the pressure release   Task Completed ☐
    button against some object on the car to relieve the pressure system.

    WARNING: *Always wear hand protection and safety goggles or glasses with side shields when
    doing this task. Avoid the hot coolant and steam.*

26. After the pressure has been relieved, shield your hands with a cloth wrapped around the flexible adapter and filler neck. Slowly turn the adapter cap from the lock position to the safety unlock position of the radiator filler neck cam. Do not detach the flexible adapter from the filler neck. Allow the pressure to dissipate completely in this safety position. Then, you can remove the adapter.

Task Completed ☐

27. If the pressure test indicated that combustion gases were escaping into the coolant, use a combustion leak detector. There are several types of these testers. Which type will you use?

_____

_____

28. Conduct the test according to the tester's manufacturer and record your results here.

_____

_____

29. Summarize your findings from these checks.

_____

_____

30. Draw some coolant out of the radiator or recovery tank with the coolant hydrometer.

Task Completed ☐

31. Check the strength of the coolant by observing the position of the tester's float. Summarize what you discovered about the coolant and what service you recommend.

_____

_____

32. Reinstall the radiator cap, and then start the engine and move the heater control to its full heat position.

Task Completed ☐

33. Turn off the engine after it has only slightly warmed up.

Task Completed ☐

34. Place a drain pan under the radiator drain plug.

Task Completed ☐

35. Make sure the engine and cooling system are not hot, and then open the drain plug. Keep the radiator cap on until the recovery tank is emptied.

Task Completed ☐

36. Once all coolant has drained, close the drain plug. Then, empty the drain pan with the coolant into the coolant recycler and process the used coolant according to the manufacturer's instructions.

Task Completed ☐

37. Connect the flushing machine to the cooling system.

Task Completed ☐

38. Follow the manufacturer's instructions for machine operation. Typically, flushing continues until clear water flows out of the system. Once flushing is complete, disconnect the flushing machine.

Task Completed ☐

39. Look up the capacity of the cooling system in the service manual. What is it?

_____

40. Prepare to put in a mixture of 50% coolant and 50% water.   Task Completed ☐

41. Locate the engine's cooling system bleed valve. Where is it?

_____

42. Loosen the bleed valve.   Task Completed ☐

43. Pour an amount of coolant that is equal to half of the cooling system's capacity into the system.   Task Completed ☐

44. Add water until some coolant begins to leak from the bleed valve. Then, close the valve.   Task Completed ☐

45. Leave the radiator cap off and start the engine.   Task Completed ☐

46. Continue to add water to the system as the engine warms up.   Task Completed ☐

47. Once the system appears full, install the radiator cap.   Task Completed ☐

48. Observe the system for leaks and watch the activity in the recovery tank. If no coolant moves to the tank, the system may need to be bled.   Task Completed ☐

Problems Encountered

_____

_____

_____

Instructor's Comments

_____

_____

_____

ENGINE REPAIR JOB SHEET 43

Identifying Cause of Overheating

Name _____ Station _____ Date _____

NATEF Correlation

This Job Sheet addresses the following **AST/MAST** task:

D.2. Identify causes of engine overheating.

Objective

Upon completion of this job sheet, you will be able to verify the operating temperature of an engine and diagnose the cause of abnormal temperatures.

Tools and Materials

Thermometer   Duct tape

Pyrometer   Clean cloth

Cooling system tester

Protective Clothing

Goggles or safety glasses with side shields

Describe the vehicle being worked on:

Year _____ Make _____ Model _____

VIN _____ Engine type and size _____

PROCEDURE

1. Engines are designed to run within a narrow range of temperatures. When they operate within that range, they are running efficiently. Typically, abnormal temperatures are only noticed by the owner when there is insufficient heat inside the vehicle during cold weather or when steam rolls out from under the hood. Why would an engine that is running at lower than normal temperatures be less efficient?

   _____

   _____

   _____

   _____

2. Name two things that may cause an engine to operate at lower than normal temperatures.

   _____

   _____

3. Name the part of the cooling system that has the responsibility for maintaining and regulating the temperature of the coolant.

   _____

   _____

4. Higher than normal operating temperatures will also affect the engine's efficiency. Name six problems that may cause the engine to run hot.

   _____

   _____

   _____

   _____

5. There are times when the engine seems to be running colder or hotter than normal but the temperature gauge reads normally. In these cases, the engine's coolant temperature should be measured and compared to the reading on the vehicle's temperature gauge. The temperature of an engine should also be verified when the activity of the powertrain computer seems to indicate that it is responding to a temperature that is different than what the temperature gauge and/or engine coolant temperature sensor indicates. Engine temperature is commonly measured with a pyrometer or a thermometer. Before measuring, look up the temperature specification for the engine's thermostat. The specification is:

   _____

   The source of the specification was: _____

6. Tape the temperature-sensing bulb of the thermometer to the upper radiator hose. Operate the engine for 15 minutes at idle speed. Now read the temperature indicated on the thermometer. The reading is:

   _____

7. Compare the reading with the specification and state the difference. Explain why it may be different.

   _____

   _____

8. Wipe off the radiator filler neck and inspect it. Describe the condition of the sealing seat and filler neck. Is it free of accumulated dirt, nicks, or anything that might prevent a good seal? Are the cams on the neck's out-turned flange bent or worn?

   _____

   _____

9. To test the cooling system for external leaks, attach a cooling system tester to the appropriate flexible adapter and carefully pump up the pressure. Pump the pressure up to the rating of the radiator cap. What is that rating? _____

10. Looking at the tester's gauge, bring the pressure up to the proper cooling system test point indicated on the dial face. Does the pressure begin to drop?

☐ Yes ☐ No

If so, check all external connections, including hoses, gaskets, and the heater core for leaks. Describe what you found.

_____

_____

11. Check all points closely for small pinhole leaks or potentially dangerous weak points in the system. Describe what you found.

_____

_____

Problems Encountered

_____

_____

_____

Instructor's Comments

_____

_____

_____

ENGINE REPAIR JOB SHEET 44

Inspect, Replace, and Adjust Drive Belts and Pulleys

Name _____ Station _____ Date _____

NATEF Correlation

This Job Sheet addresses the following **MLR** task:

C.2. Inspect, replace, and adjust drive belts, tensioners, and pulleys; check pulley and belt alignment.

This Job Sheet addresses the following **AST/MAST** task:

D.3. Inspect, replace, and adjust drive belts, tensioners, and pulleys; check pulley and belt alignment.

Objective

Upon completion of this job sheet, you will be able to inspect, replace, and adjust drive belts, tensioners, and pulleys.

Tools and Materials

Hand tools

Belt tension gauge

Pry bar

Service information

Protective Clothing

Goggles or safety glasses with side shields

Describe the vehicle being worked on:

Year _____ Make _____ Model _____

VIN _____ Engine type and size _____

Describe general condition:

_____

PROCEDURE

Serpentine Belts

1. Check the condition of the drive belt. Carefully look to see if it is cracked or glazed. What is the condition of the belt?

   _____

   _____

2. If the belt needs to be replaced, disconnect the electric cooling fan at the radiator, if the vehicle has one.   Task Completed ☐

3. Always use the exact replacement size of belt. The size of a new belt is typically given, along with the part number, on the belt container. What is the part number?

   _____

4. Locate a belt routing diagram in a service manual or on an under-hood decal. Where did you find this diagram?

   _____

5. Compare the diagram with the routing of the old belt. If the actual routing is different from the diagram, draw the existing routing here.

6. Loosen the tension on the old belt and remove the belt. How did you relieve the tension?

   _____

   _____

7. Inspect the grooves of the drive pulleys for rust, oil, wear, and other damage. What did you find and what service do you recommend?

   _____

   _____

8. Check the alignment of the pulleys. What service do you recommend?

   _____

   _____

9. Make sure the belt tensioner or idler pulley is working properly. This pulley may be a spring-loaded tensioner or an adjustable pulley. Describe its condition.

   _____

   _____

10. After all the parts are determined to be in good order, install the new belt. Make sure to wrap it according to instructions. Also, make sure the ribs of the belt are seated in the matched grooves on the pulleys.

    Task Completed ☐

11. Once the belt is fully routed, put tension on the belt and adjust it to specifications. What are the specifications?

    _____

    _____

Problems Encountered

_____

_____

_____

Instructor's Comments

_____

_____

_____

ENGINE REPAIR JOB SHEET 45

Servicing a Water Pump

Name _____ Station _____ Date _____

NATEF Correlation

This Job Sheet addresses the following **AST/MAST** task:

D.5. Inspect, test, remove, and replace water pump.

Objective

Upon completion of this job sheet, you will be able to inspect, test, remove, and replace a water pump.

Tools and Materials

Stethoscope

Basic hand tools

Protective Clothing

Goggles or safety glasses with side shields

Describe the vehicle being worked on:

Year _____ Make _____ Model _____

VIN _____ Engine type and size _____

PROCEDURE

NOTE: *Some water pumps are driven by the timing belts and are not visual without removing the timing covers. Consult the service information for the procedures to inspect and replace the water pump.*

1. Begin your inspection of the water pump by looking for evidence of leaks. Look carefully for signs of leakage at the weep hole in the water pump casting. If there is a leak at the weep hole, what is indicated? What did you find during your general inspection of the water pump?

   _____

   _____

2. If you suspect that the water pump is related to a noise problem or if it appears that the water pump seal is leaking, check for the following (if applicable). Record your findings next to each item.

   a. A bent fan.

   _____

   _____

    b. A piece of the fan is missing.

    _____

    c. A cracked fan blade.

    _____

    d. Fan mounting surfaces that are not clean or flush.

    _____

    e. A worn fan clutch.

    _____

3. To check the water pump, start the engine and listen for a bad bearing, using a mechanic's stethoscope. Carefully avoiding moving parts, place the stethoscope on the bearing or pump shaft. Describe what you heard and what this indicates.

    _____

    _____

    _____

4. Turn the engine off and remove the fan belt and shroud (if applicable). Grasp the fan and attempt to move it in and out and up and down. How much were you able to move it and what is indicated by this?

    _____

    _____

    _____

5. Reinstall the fan belt and shroud. Make sure the belt is tightened to the correct tension.

    Task Completed ☐

6. Now, warm up the engine and run it at idle speed. Squeeze the upper hose connection with one hand and accelerate the engine with the other hand. What do you feel and why do you feel it?

    _____

    _____

7. If the water pump is defective or if it leaks, it must be replaced. Begin replacement by draining the coolant from the cooling system.

    Task Completed ☐

8. Make sure you catch all the coolant as it drains and recycle it as directed by local regulations.

    Task Completed ☐

9. Remove all components, such as the drive belts, fan, fan shroud, shaft spacers, or viscous drive clutch, that block accessibility to the water pump. What do you need to remove in order to remove the water pump?

    _____

    _____

10. Loosen and remove the bolts in a crisscross pattern from the center outward. Insert a rag into the block opening and scrape off any remains of the old gasket.

Task Completed ☐

11. When replacing a water pump, always follow the procedures recommended by the manufacturer. What are those recommendations?

_____

_____

_____

12. What should be done to the gasket and sealing surfaces prior to installation of the water pump?

_____

_____

13. Install the mounting bolts and tighten them evenly in a staggered sequence to the torque specifications. What are those specifications?

_____

14. Check the pump to make sure it rotates freely. What happens when you rotate the pump?

_____

_____

15. Reinstall everything you had to remove to gain access to the water pump. What components did you need to install?

_____

_____

_____

16. Make sure the drive belt(s) are tightened to the proper tension. What is the required tension for each belt?

_____

17. Refill the cooling system to the proper level.

Task Completed ☐

18. Pressure check the system and look for signs of leakage.

Task Completed ☐

19. If no leaks are evident, start the engine and allow it to run until it reaches normal operating temperature. Then, bleed the air from the system. How did you do that?

_____

_____

20. Shut the engine off and look for leaks, then replenish the cooling system if necessary.

Task Completed ☐

Problems Encountered

_____

_____

_____

Instructor's Comments

_____

_____

_____

ENGINE REPAIR JOB SHEET 46

Remove and Replace a Radiator

Name _____ Station _____ Date _____

NATEF Correlation

This Job Sheet addresses the following **AST/MAST** task:

D.6. Remove and replace radiator.

Objective

Upon completion of this job sheet, you will be able to remove and replace a radiator.

Tools and Materials

Drain pan

Protective Clothing

Goggles or safety glasses with side shields

Describe the vehicle being worked on:

Year _____ Make _____ Model _____

VIN _____ Engine type and size _____

PROCEDURE

1. Disconnect the negative terminal of the battery.   Task Completed ☐

2. Drain the cooling system and recycle or dispose of the coolant according   Task Completed ☐
   to local regulations.

3. Loosen and remove the hose clamps for the hoses that connect to the   Task Completed ☐
   radiator.

4. Disconnect the transmission cooler (if so equipped) lines and plug them.
   What did you use to plug them?

   _____

5. Disconnect the wiring harness connector to the electric cooling fan, if so   Task Completed ☐
   equipped.

6. Disconnect any sensor wires that may be attached to the radiator. What
   sensors are installed in the radiator?

   _____

   _____

7. Remove the fasteners that attach the cooling fan assembly to the radiator. Or remove the attaching bolts for the fan shrouding. Which did you need to remove?

_____

_____

8. Remove the cooling fan assembly or shroud. What else did you need to remove?

_____

_____

9. Check the service information to see if the air conditioning condenser must be removed with the radiator. If so, what must you do to disconnect the condenser from the A/C system?

_____

_____

_____

_____

10. Remove the upper radiator mounts.   Task Completed ☐

11. Remove the radiator.   Task Completed ☐

12. Place the new or rebuilt radiator into its lower mounts. Make sure the   Task Completed ☐
    rubber insulators are in place after doing this.

13. Install and bolt the upper mounts.   Task Completed ☐

14. If the A/C condenser was disconnected, reconnect it according to the   Task Completed ☐
    manufacturer's recommendations.

15. Install all parts of the cooling fan assembly or shroud.   Task Completed ☐

16. Connect any sensor wires that were disconnected during the removal of   Task Completed ☐
    the radiator.

17. Connect the electrical connector for the cooling fan.   Task Completed ☐

18. Unplug the transmission cooler fittings and attach the lines to the radiator. Task Completed ☐

19. Inspect the upper and lower radiator hoses. If they are in good shape,
    reuse them. If they are questionable or bad, replace them. Describe the
    condition of the hoses.

_____

_____

20. With new clamps, attach the upper and lower radiator hoses to the radiator. Task Completed ☐

21. Refill the cooling system with coolant. How much coolant did you need to add?

_____

_____

22. Pressure check the system and look for evidence of leaks. Describe the results of this check.

_____

_____

23. If there are no leaks, connect the battery.   Task Completed ☐

24. Start the engine and look for leaks.   Task Completed ☐

25. Allow the engine to reach normal operating temperature, and then bleed any air that was trapped in the cooling system.   Task Completed ☐

26. Check and replenish the transmission fluid level.   Task Completed ☐

27. Turn off the engine and recheck the coolant level.   Task Completed ☐

Problems Encountered

_____

_____

_____

Instructor's Comments

_____

_____

_____

ENGINE REPAIR JOB SHEET 47

Servicing a Thermostat

Name _____ Station _____ Date _____

NATEF Correlation

This Job Sheet addresses the following **MLR** task:

C.3. Inspect, test, and replace thermostat and gasket/seal.

This Job Sheet addresses the following **AST/MAST** task:

D.7. Inspect, test, and replace thermostat and gasket/seal.

Objective

Upon completion of this job sheet, you will be able to remove, inspect, and test the thermostat, bypass, and housing, and then properly reinstall the assembly.

Tools and Materials

Thermometer

Container for water

Heat source for heating the water

Protective Clothing

Goggles or safety glasses with side shields

Describe the vehicle being worked on:

Year _____ Make _____ Model _____

VIN _____ Engine type and size _____

PROCEDURE

1. Where is the thermostat located?

   _____

   _____

2. Thoroughly inspect the area around the thermostat and its housing. Describe the result of that inspection.

   _____

   _____

   _____

3. There are several ways to test the opening temperature of a thermostat. The first method does not require that the thermostat be removed from the engine. Begin by removing the radiator pressure cap from a cool radiator and insert a thermometer into the coolant.

Task Completed ☐

4. Start the engine and let it warm up. Watch the thermometer and the surface of the coolant. When the coolant begins to flow or move in the radiator, what is indicated?

_____

_____

5. When the fluid begins to flow, record the reading on the thermometer. If the engine is cold and coolant circulates, this indicates the thermostat is stuck open and must be replaced. The measured opening temperature of the thermostat was:

_____

6. Compare the measured opening temperature with the specifications. The specifications call for the thermostat to open at what temperature? _____ What do you recommend based on the results of this test?

_____

_____

7. The other way to test a thermostat is to remove it. Begin removal by opening the radiator pressure cap to relieve pressure, if the cap is still on.

Task Completed ☐

8. Drain coolant from the radiator until the level of coolant is below the thermostat housing. Recycle the coolant according to local regulations. What did you do with the drained coolant?

_____

_____

_____

9. Disconnect the upper radiator hose from the thermostat housing.

Task Completed ☐

10. Unbolt and remove the housing from the engine. The thermostat may come off with the housing.

Task Completed ☐

11. Thoroughly clean the gasket surfaces for the housing. Make sure the surfaces are not damaged while doing so and that gasket pieces do not fall into the engine at the thermostat bore.

Task Completed ☐

12. Carefully inspect the thermostat housing. Describe your findings.

_____

_____

13. Suspend the thermostat while it is completely submerged in a small container of water. Make sure it does not touch the bottom of the container. What did you use to suspend it?

_____

_____

_____

14. Place a thermometer in the water so that it does not touch the container and only measures water temperature.

Task Completed ☐

15. Heat the water. When the thermostat valve barely begins to open, read the thermometer. What was the measured opening temperature of the thermostat? Compare that with the specifications.

_____

_____

16. Remove the thermostat from the water and observe the valve. Record what happened and your conclusions about the thermostat.

_____

_____

_____

17. Carefully look over the new (or old if okay) thermostat. Identify which end of the thermostat should face toward the radiator. How did you determine the proper direction for installation?

_____

_____

18. Fit the thermostat in the recessed area in the engine or thermostat housing. Where was the recess?

_____

19. Refer to the installation instructions on the gasket's or seal's container. Should an adhesive and/or sealant be used with the gasket? If so, what?

_____

_____

20. Install the gasket/seal according to the instructions.

Task Completed ☐

21. Install the thermostat housing. Before tightening it in place, make sure it is fully seated and flush onto the engine. Failure to do this will result in a broken housing.

Task Completed ☐

22. Tighten the bolts evenly and carefully and to the correct specifications. What are the specifications?

_____

23. Connect the upper radiator hose to the housing with a new clamp. Why should you use a new clamp?

_____

_____

_____

24. Pressurize the system and check for leaks. Why should you do this now?

_____

_____

25. Replenish the coolant and bring it to its proper level. Install the radiator cap.   Task Completed ☐

26. Run the engine until it is at a normal operating temperature. Did the cooling system warm up properly and does it seem that the thermostat is working properly?

_____

_____

27. Recheck the coolant level.   Task Completed ☐

Problems Encountered

_____

_____

_____

Instructor's Comments

_____

_____

_____

ENGINE REPAIR JOB SHEET 48

Clean, Inspect, Test, and Replace Cooling Fans, and Cooling System-Related Temperature Sensors

Name _____ Station _____ Date _____

NATEF Correlation

This Job Sheet addresses the following **AST/MAST** task:

D.8. Inspect and test fan(s) (electrical or mechanical), fan clutch, fan shroud, and air dams.

Objective

Upon completion of this job sheet, you will be able to inspect and test electrical or mechanical cooling fan(s), a fan clutch, fan shroud, and air dams.

Tools and Materials

Hand tools

Test light (circuit tester)

DMM

Jumper wire

Service information

Protective Clothing

Goggles or safety glasses with side shields

Describe the vehicle being worked on:

Year _____ Make _____ Model _____

VIN _____ Engine type and size _____

Describe general condition:

_____

PROCEDURE

Electric Fans

WARNING: *The electric cooling fans can operate with the engine and the key both turned off. The fan can come on at any time if the engine is hot or there is a problem with a sensor. Keep your hands away from the fan at all times.*

1. Visually inspect the electric cooling fan(s) and related wiring to determine if they are clean and properly connected.   Task Completed ☐

2. Run the engine until it reaches normal operating temperature to determine if the electric cooling fan is working. (On some vehicles, the cooling fan can be commanded on by a scan tool.)   Task Completed ☐

3. If the fan is operating properly, clean the fan and surrounding area with compressed air.   Task Completed ☐

4. If the fan is not working, check the power wire to the fan to determine if there is power.   Task Completed ☐

5. Check the ground circuit to determine if the system has a good ground.   Task Completed ☐

6. If power and a good ground are present, remove the electric fan assembly and replace it.   Task Completed ☐

7. If there is no power, check the fuse and fan relay. The location of the fuse and relay can be found in the service manual.   Task Completed ☐

8. Replace the bad components, and then retest the fan operation.   Task Completed ☐

9. If there is not a good ground, check the temperature sensor that controls the electric fan. The location of the sensor and the proper test procedure can be found in the service manual.   Task Completed ☐

10. If the sensor fails the test, replace it. Check the operation of the electric fan.   Task Completed ☐

Mechanical Fans

1. Inspect the fan blades for stress cracks and damaged blades. Describe what you found.

   _____

   _____

2. Inspect the drive belt. Describe what you found.

   _____

   _____

3. If the vehicle has a fan clutch, inspect it for signs of fluid leakage. Describe what you found and state what this indicates.

   _____

   _____

4. Move the fan blades attached to the fan clutch to check its integrity. Describe what you found and state what this indicates.

   _____

   _____

Fan Shrouds

1. Inspect the fan shroud and its attachments. Describe the condition.

   _____

   _____

2. Why is it important to have the shroud intact and in good condition?

   _____

   _____

Problems Encountered

_____

_____

_____

Instructor's Comments

_____

_____

_____

ENGINE REPAIR JOB SHEET 49

Measure Engine Oil Pressure

Name _____ Station _____ Date _____

NATEF Correlation

This Job Sheet addresses the following **AST/MAST** task:

D.9. Perform oil pressure tests; determine necessary action.

Objective

Upon completion of this job sheet, you will be able to perform oil pressure tests on an engine.

Tools and Materials

Adapter fitting   Open-end wrench

Fender covers   Service information

Oil-pressure test gauge   Tachometer

Protective Clothing

Safety goggles or glasses with side shields

Describe the vehicle being worked on:

Year _____ Make _____ Model _____

VIN _____ Engine type and size _____

Describe general condition:

_____

PROCEDURE

1. Look up the oil pressure specifications for this engine in the appropriate service information:

    Oil pressure _____

    Engine speed _____

2. Locate the oil-pressure sending unit; usually it is on the lower side of the engine block. Disconnect the wire from the sending unit and use an open-end wrench to remove the sender.   Task Completed ☐

3. Tighten the oil-pressure test gauge into the hole in the block where the sender was removed. Use an adapter fitting, if necessary, to make the connection.   Task Completed ☐

4. Check the engine's oil level and fill, if required.   Task Completed ☐

> **WARNING:** *Be extremely careful when working near a running engine. Always wear safety goggles or glasses with side shields when working around moving machinery and be sure your clothing is not loose.*

5. Start the engine and observe the pressure reading on the gauge. Make sure the engine speed is set to specifications for testing pressure. If necessary, use a tachometer and adjust the engine idle speed. Record the measured oil pressure, and then turn the engine off.   Task Completed ☐

6. Is your measurement below specifications?

   _____

7. Consult the appropriate service information and list your conclusions and recommendations.

   _____

   _____

8. Remove the test gauge and adapter fitting. Reinstall the oil pressure sender and connect the wire. Start the engine and check for leaks.   Task Completed ☐

Problems Encountered

_____

_____

_____

Instructor's Comments

_____

_____

_____

ENGINE REPAIR JOB SHEET 50

Perform Oil and Filter Change

Name _____ Station _____ Date _____

NATEF Correlation

This Job Sheet addresses the following **MLR** task:

C.5. Perform oil and filter change.

This Job Sheet addresses the following **AST/MAST** task:

D.10. Perform oil and filter change.

Objective

Upon completion of this job sheet, you will be able to properly change an engine's oil and oil filter.

Tools and Materials
Rags
Funnel
Oil filter wrench
Lift

Protective Clothing
Goggles or safety glasses with side shields

Describe the vehicle being worked on:
Year _____ Make _____ Model _____

VIN _____ Engine type and size _____

PROCEDURE

1. Always make sure the vehicle is positioned safely on a lift or supported by jack stands. Before raising the vehicle, allow the engine to run a while. After it is warm, turn off the engine.

   Task Completed ☐

2. Place the oil drain pan under the drain plug before beginning to drain the oil.

   Task Completed ☐

3. Loosen the drain plug with the appropriate wrench. After the drain plug is loosened, quickly remove it so the oil can freely drain from the oil pan. Make sure the drain pan is positioned so it can catch all of the oil. Describe the color and condition of the oil.

   _____

   _____

4. While the oil is draining, use an oil filter wrench to loosen and remove the oil filter. Describe what you needed to do in order to do this.

_____

_____

5. Make sure the oil filter seal came off with the filter. Then, place the filter into the drain pan so it can drain. After it has completely drained, discard the filter according to local regulations. What are your local requirements for disposing of oil filters and oil?

_____

_____

6. Wipe off the oil filter sealing area on the engine block. Then, apply a coat of clean engine oil onto the new filter's seal.   Task Completed ☐

7. Install the new filter and hand tighten it. Oil filters should be tightened according to the directions given on the filter. What are the instructions?

_____

_____

8. Prior to installing the drain plug, wipe off its threads and sealing surface with a clean rag. Inspect the threads and describe their condition. Replace the plug's sealing washer if equipped.

_____

_____

9. The drain plug should be tightened according to the manufacturer's recommendations. Over-tightening can cause thread damage, while under-tightening can cause an oil leak. What is the tightening spec for the plug?

_____

10. With the oil filter and drain plug installed, lower the vehicle and remove the oil filler cap.   Task Completed ☐

11. Carefully pour the oil into the engine. The use of a funnel usually keeps oil from spilling onto the engine. What type of oil is recommended and how much?

_____

12. After the recommended amount of oil has been put into the engine, check the oil level.   Task Completed ☐

13. Start the engine and allow it to reach normal operating temperature. While the engine is running, check the engine for oil leaks, especially around the oil filter and drain plug. If there is a leak, shut down the engine and correct the problem. Were there any leaks? Where?

_____

14. After the engine has been turned off, recheck the oil level and correct it as necessary.   Task Completed ☐

Problems Encountered

_____

_____

_____

Instructor's Comments

_____

_____

_____

ENGINE REPAIR JOB SHEET 51

Service Auxiliary Oil Coolers

Name _____ Station _____ Date _____

NATEF Correlation

This Job Sheet addresses the following **AST/MAST** task:

D.11. Inspect auxiliary coolers; determine necessary action.

Objective

Upon completion of this job sheet, you will be able to inspect and replace auxiliary oil coolers.

Tools and Materials

OSHA-approved air nozzle	Bucket of water
Vacuum pump	Drain pan
Line plugs	Clean rags

Protective Clothing

Goggles or safety glasses with side shields

Describe the vehicle being worked on:

Year _____ Make _____ Model _____

VIN _____ Engine type and size _____

PROCEDURE

1. Where is the auxiliary oil cooler located?

   _____

   _____

2. Carefully check the lines connecting the cooler to the engine. Are there signs of leakage or damage? Summarize your findings.

   _____

   _____

3. Carefully check the cooler. Are there signs of leakage or damage? Summarize your findings.

   _____

   _____

4. Place a drain pan under the cooler.   Task Completed ☐

5. Disconnect the cooler lines and remove the cooler.   Task Completed ☐

6. To leak test the cooler, a vacuum pump or shop air can be used. Using a line plug, close off the outlet fitting of the cooler.

Task Completed ☐

7. Connect the vacuum pump line to the inlet fitting at the cooler.

Task Completed ☐

8. Run the vacuum pump. If there is a leak in the cooler, vacuum will not build. What did you observe?

_____

_____

9. Insert the cooler into a bucket of water, making sure water does not enter into it.

Task Completed ☐

10. Insert the air nozzle into the inlet fitting at the cooler.

Task Completed ☐

11. Release the air pressure into the cooler while observing the water in the bucket. If there are bubbles, the cooler leaks. What did you observe?

_____

_____

12. If the oil cooler checks out fine, remount it to the vehicle.

Task Completed ☐

13. Using new O-rings and/or seals, connect the lines to the cooler.

Task Completed ☐

14. Look up the capacity of the cooler and add that much oil to the engine.

Task Completed ☐

15. Start the engine and check for leaks.

Task Completed ☐

16. After the engine is warmed up, turn off the engine and recheck the engine's oil level.

Task Completed ☐

Problems Encountered

_____

_____

_____

Instructor's Comments

_____

_____

_____

ENGINE REPAIR JOB SHEET 52

Servicing Oil Pressure and Temperature Sensors

Name _____ Station _____ Date _____

NATEF Correlation

This Job Sheet addresses the following **AST/MAST** task:

D.12. Inspect, test, and replace oil temperature and pressure switches and sensors.

Objective

Upon completion of this job sheet, you will be able to inspect, test, and replace oil temperature and pressure switches and sensors.

Tools and Materials

Ohmmeter

Oil pressure tester

Protective Clothing

Goggles or safety glasses with side shields

Describe the vehicle being worked on:

Year _____ Make _____ Model _____

VIN _____ Engine type and size _____

PROCEDURE

1. Describe the type(s) of oil gauges and/or warning lights the engine is equipped with.

   _____

   _____

   _____

2. Explain the purpose of each.

   _____

   _____

   _____

3. Locate the specifications for each in the service information and record the specifications here.

   _____

   _____

   _____

4. When you turn the ignition on with the engine off, do the indicator lamps light? What are the readings on the gauges?

   _____

   _____

5. Start the engine. Do the lamps turn off? Do the readings on the gauges change?

   _____

   _____

6. What can you conclude so far?

   _____

   _____

7. With the engine off, carefully examine the area around each of the sensors. Look for signs of oil leakage and record your findings.

   _____

   _____

8. Also look for oil inside the protective boots for the electrical connectors. What can you conclude?

   _____

   _____

9. Disconnect the electrical connector to each of the sensors and connect an ohmmeter from the terminal of the sensor to ground. Compare the reading to the specifications. What can you conclude from this test?

   _____

   _____

10. Start the engine and look at the ohmmeter reading. Compare the reading to the specifications. What can you conclude from this test?

    _____

    _____

11. If the oil pressure sensor is suspected of being bad, note the oil pressure reading shown on the oil pressure gauge on the instrument panel while the engine is idling.

_____

_____

12. Turn the engine off and remove the sensor. Connect an oil pressure tester to the sensor's bore. Start the engine and allow it to idle. Record the oil pressure shown on the test gauge.

_____

13. Compare the reading on the test gauge with the reading taken at the instrument panel. What can you conclude?

_____

_____

14. Install the sensor or a new one and reconnect the electrical connector to it. What sensor did you install?

_____

_____

15. If the sensor(s) tested fine but there is still misinformation at the gauges or indicator lamps, the gauge or lamp circuit must be tested. Is it necessary to do this on this vehicle?

_____

_____

_____

_____

Problems Encountered

_____

_____

_____

Instructor's Comments

_____

_____

_____

ENGINE REPAIR JOB SHEET 53

Service and Install Oil Pump and Oil Pan

Name _____ Station _____ Date _____

NATEF Correlation

This Job Sheet addresses the following **MAST** task:

D.13. Inspect oil pump gears or rotors, housing, pressure relief devices, and pump drive; perform necessary action.

Objective

Upon completion of this job sheet, you will be able to inspect and replace the pans, covers, gaskets, and seals, as well as inspect the oil pump gears or rotors, housing, pressure relief devices, and pump drive.

Tools and Materials

Feeler gauge   Service information

Measuring scale   Sockets and ratchet

Outside micrometer   Straightedge

Pan gaskets   Torque wrench

Protective Clothing

Goggles or safety glasses with side shields

Describe the engine being worked on:

Year _____ Make _____ Model _____

VIN _____ Engine type and size _____

Describe the type of oil pump:

_____

PROCEDURE

1. Using the service information, locate the required specifications for the Report Sheet for Oil Pump Service, found at the end of this job sheet, and enter them there.   Task Completed ☐

   WARNING: *Always wear safety goggles or glasses with side shields when doing this task. The oil pressure relief valve is controlled by spring pressure, and the retainer may fly out of the housing, causing severe injury.*

2. Disassemble the oil pump and relief valve assembly in clean solvent and allow them to dry.   Task Completed ☐

3. Place a straightedge across the pump cover, and try to push a 0.002-in. feeler gauge between the cover and straightedge. If the gauge fits, the cover is worn. Record your results on the Report Sheet for Oil Pump Service.

Task Completed ☐

4. Measure the thickness and diameter of the rotors or gears. Compare the measurements with specifications. Measurements smaller than the specifications mean the pump must be replaced. Record your results on the Report Sheet for Oil Pump Service.

Task Completed ☐

5. Assemble the gears or rotors into the pump body. Measure the clearance between them with a feeler gauge. Record your results on the Report Sheet for Oil Pump Service. If the clearance is larger than specifications, replace the pump.

Task Completed ☐

6. Position a straightedge across the gears or rotors, and measure the clearance with a feeler gauge. If the measurement is larger than the specifications, replace the pump. Record your results on the Report Sheet for Oil Pump Service.

Task Completed ☐

7. Measure the relief valve spring height and record your findings on the Report Sheet for Oil Pump Service.

Task Completed ☐

8. Lubricate all pump parts and reassemble the pump. Tighten all bolts to specifications.

Task Completed ☐

9. Install the pump drive or extension and install the pump on the engine. Torque the mounting bolts to specifications.

Task Completed ☐

10. Inspect the sealing surface of the pan and cylinder block. Install new gaskets and seals on the oil pan. Use gasket sealer or sealant, as specified. Install the pan and tighten the pan bolts to specifications.

Task Completed ☐

11. Install a new pan drain-plug gasket and install the drain plug. Tighten to specifications.

Task Completed ☐

Problems Encountered

_____

_____

_____

Instructor's Comments

_____

_____

_____

REPORT SHEET FOR OIL PUMP SERVICE

Measuring	Specifications	Actual
Cover warpage		
Rotor diameter		
Rotor thickness		
Rotor clearance (backlash)		
Gear (rotor)-to-cover clearance		
Relief spring free length		
		Torque Specifications
Oil pump cover bolts		
Oil pump mounting bolts		
Oil pan bolts		
Oil pan drain plug		

Conclusions and Recommendations _____

_____

_____

— NOTES —